AF225676

SILENT TEARS

The Life and Times of a Nuclear Test Cloud-Chaser

JOHN FOLKES

This edition first published in paperback by
Michael Terence Publishing in 2022
www.mtp.agency

Copyright © 2022 John Folkes

John Folkes has asserted the right to be identified as
the author of this work in accordance with the
Copyright, Designs and Patents Act 1988

ISBN 9781800944572

No part of this publication may be reproduced, stored
in a retrieval system, or transmitted, in any form or
by any means, electronic, mechanical, photocopying,
recording or otherwise, without the prior
permission of the publisher

Cover design
Copyright © 2022 Michael Terence Publishing

Michael Terence
Publishing

It is to my fellow veterans and to those who did not survive putting Service needs before their own safety during the 1950s and 1960s British nuclear tests, that I dedicate this book. May their unselfish service and in many cases sacrifice for their Country soon be recognised with gratitude from the British government.

Notice For
First Peoples of Australia

Aboriginal and Torres Strait Islander peoples are advised that this publication contains images of people from an Aboriginal community of South Australia taken in 1956 who may have since passed away.

A list of pictures is included (without images) along with page numbers to help those who wish to avoid seeing these images.

Preface

This is a story born out of personal experience as a nuclear cloud flyer during the 1956 British nuclear tests in South Australia at a testing site in a remote desert area called Maralinga.

The blueprint for my life in the Royal Air Force began at the age of fifteen and a half when I joined what was then called the Boy Entrant Service. This involved a commitment to eighteen months of strict basic training in airmanship and as a trainee aircraft mechanic. This required both self-discipline and self-reliance that were to play an essential part in my earlier life, and would be put to the test. These qualities, matched with a strong sense of duty towards service life, prepared me for future demands that were to come from a direction that I then never thought possible.

CONTENTS

PICTURES

CHAPTER 1

The Formative Years

I begin my story at the outbreak of the Second World War. From the early age of five, my formative years were to be constantly disrupted as my family and I moved frequently, either for my parents' employment or because of the heavy bombing by the German Luftwaffe. This disruptive lifestyle was to play havoc with my schooling. When the bombing intensified, my parents decided to move to Bath in Somerset. Shortly after, on the 25th April, 1942, German air raids switched over to Bath and civilian targets were chosen for their cultural and historical reasons rather than their strategic or military value. My memories of this period are a bit sketchy as I was at this time about six or seven and do not recall having attended any school. I do remember there being a large stream at the bottom of the garden on which I floated down on an old wooden door that I used as a raft. As the raids got worse, and not having an air raid shelter we would frequently have to leave the house and head to the open fields.

I vividly remember the close call we had one evening. We had to leave the house during a heavy raid and it was decided that our best option would be to gather what belongings we could carry and head for a place on the nearby hill, as I recall it was known as Peas Down. On our way we passed under a large railway arch where many people were gathered for protection. Hidden under a cloak of blankets, and protectively squeezed between Mum and Dad, we walked up the steep road with houses either side severely damaged and engulfed in flames. About half way up we felt a blast from behind us and on looking back saw the railway arch, which moments before we had passed under, now smothered in smoke and flames reduced to rubble. No one could have survived. What we were to witness on reaching the top of the hill was the remarkable site of several German Ju 88 bombers passing over us, then releasing their bomb loads over what I was told was the grounds of the Bath football Club situated in the middle of a residential area. What was striking was the sight of the long canopy profile of the Ju 88s being fully illuminated where their aircrew could be easily seen. Having been so lucky to have survived, on return to where we lived the house was severely damaged. It was shortly after this the decision was taken for us to return to Sydenham in South East London and take our chances there.

During this time my much older brother, Malcolm, who was serving in the Royal Navy on Motor Torpedo Boats (MTBs) would apply for shore leave whenever possible. To celebrate his infrequent short periods of leave, Mum and Dad would organize family get-togethers in the local pub, The Woodman, not far from home where we lived at 18 Mount Ash Road in Sydenham, Forest Hill, South East London. If on one of those occasions we heard the chilling sound of the air raid siren, we would all retreat to the relative safety of the cellar below where there was a temporary bar and a piano. This was fortunate because Malc, as he was known to us, was an accomplished self-taught jazz pianist. He played in the style of Harry Roy whose band he toured with were popular at the time and called the Tiger Ragamuffins, with one of their then favourite tunes being 'Tiger Rag'. Accompanied at the keyboard by Bert, a close friend of the family, they would soon have everyone singing and dancing while playing their hearts out on the old upright piano, which was topped with a line of glasses spilling beer as it rocked backward and forwards. This served to drown out the horrors going on above, with the only visible evidence of the air raid being that of dust falling from the ceiling from the vibration of the explosions. I remember Mum used to say that if we have a direct hit we might as well die happy. Somehow, we survived the rest of the war, between staying in the house and the Anderson shelter in the back garden. My Primary education in the remaining war years continued to be a hit and miss affair.

Just after the war, in the latter part of 1947, my adopted parents took up service employment to a Mr Henry Price, a Conservative candidate for West Lewisham who later was to become an MP and the Minister for Fuel and Power under Winston Churchill. This move was to coincide with my transition from Primary to Secondary education, which was to bring into sharp focus the serious shortcomings in my educational standard and soon to give me serious problems that made me realise how far I was behind my peers. This situation was highlighted on my first day at my new school where I encountered the word "maths" listed every day on the notice board timetable, and which I thought was something to do with religious instruction.

In those days it was my experience that the war years were no excuse for having lack of knowledge, so punishment by cane was much the order of the day. An example of this was that it was common practice to have a maths test every Friday morning, followed shortly after the lunch time break by the results being read out. The names of those with the highest marks were read out first, in a line of descending marks following around the room. In my case, I was always last, until one week when we were shown how to calculate the quantity of areas in a way that I was able to follow without difficulty. On the following Friday afternoon for the first time, I felt confident that I had done well with the test. All the results were read out except for mine. As members of the class were taking their places, I thought that I was being left to last on this occasion as an example of how I had improved. My marks were eventually read out and were among the highest in the class. But to my surprise I was told that I was going to be made an example of for cheating. I was sent to the headmaster's room to fetch the cane and punishment book and in front of the class struck several times across the back of my bare legs and on both hands. My situation was made worse by the fact that individual test results were reflected in competitive house placings. I was in Ruskin house whose colour was green.

Clearly my results were a dead weight so it was not unusual for various members of my team to gang up on me. This retribution for letting the side down, as they saw it, was usually reserved for lunch time when teachers were not around, and after school, whereupon I would walk the streets, hiding until school started again or until I could find a safe way home. On the occasions when I was trapped, I would be pinned to the wall and pricked on my arms and legs with opened safety pins. The fear of English and maths haunted me throughout my schooldays, because in the eyes of everyone, including the staff, I appeared to be lazy and stupid.

Schooling for me was an ordeal and dread, but somehow, I had to see it through. However, sport was something I could compete with on my own terms and eventually I represented the school on a regular basis playing football and cricket. The walk I had each day was about two miles, during which I passed a high street shop showing an advertisement outside of men in white overalls working on bi-planes at what seemed to be Croydon aerodrome. This image never failed to attract my attention. The fact that it was an advert for Craven "A" cigarettes did not register with me at all. This billboard gave me the first seeds of thought that someday I wanted to work on aircraft, although my schooling constantly reminded me how far I was from ever becoming an aircraft engineer. One option that was open to me was to join the local Air Training Corps (ATC) and gain some experience if not some real contact with aircraft, This I duly did, joining 1475 Squadron, and I was not disappointed because we were given lessons on both German and English aircraft recognition. Better still, we were taught the basics of air navigation with the aid of a link trainer, a forerunner of today's flight simulator, in which we would communicate with Morse code, and also learned about what makes an aeroplane fly. In addition, we were taken on weekend camps sleeping under canvas at operational RAF (Royal Air Force) airfields, with air experience thrown in. I can remember how thrilled I was when I first flew in an A.V. ROE Anson taking off at RAF Kenley on a round trip over London. On another occasion, I flew in a DC3 Dakota.

As I got older, the progress in my education did not greatly improve and it became obvious that if I wanted to achieve my ambition working on aircraft, I would at some stage need to pass an entrance examination. My chosen route to a career was to join the RAF. Telling my parents of my decision was a shock to them, as life in the military was not something they had any previous experience of.

An early photograph of the author in ATC uniform at the age of 13.

The inevitable question about my educational situation came up, but once they were convinced of my genuine desire to join the RAF, they approached Mr Price with a view for him to find out what educational requirements I would have to meet to become a RAF Boy Entrant. Once he received the information, and being fully aware of the shortcomings in my education, he kindly offered his help by setting me homework tailored to pass the Boy Entrant examination. It followed that each Saturday morning he would devote a small amount of time in his study marking my work and talking me through the areas where I had made mistakes. This process was to take several months, progressing with maths and English until he was satisfied that I was ready to formally apply for the eighteen-month Boy Entrant apprenticeship. On the 3rd August, 1951, Mr Price, on my behalf, formally submitted an application with his full written recommendation for me to join the RAF Boy Entrance scheme. All I could do now was wait for the reply, which duly arrived accepting my application and the Boy Entrant training now subject to passing the entrance examination and medical to be carried out at RAF Cosford.

The day finally arrived when my school days were finished, the occasion celebrated by a prize giving ceremony. The sum total of my educational achievements was marked with the presentation of a book as a prize for work I had done in metal work.

Dear Sir

John Stuart Folkes, aged 15, has lived with me at this address for nearly four years along with his parents. I have thus had an unrivalled opportunity of observing him and assessing his character.

I am happy to be able to assure you that his moral character is extremely high.

He is honest, truthful, clean-living, and reliable.

He is very interested in sport, especially cricket and football, and has played for his school, at both.

He is quiet, modest and a good sportsman; he takes knocks and defeat exceptionally well.

He has been an extremely keen member of the local A.T.C. for some time past and I often noticed the pride he takes in his appearance.

Yours faithfully

Henry Price
Member of Parliament and Member of the London County Council for West Lewisham.

The Royal Air Force Boy Service

Autumn 1951 at the age of fifteen it was time to go to RAF Cosford for the exams, I was seen off at Waterloo Station by Dad who sent me on my way with tomato and egg sandwiches. I felt reassured as I started on my long journey to RAF Cosford, near Wolverhampton, as there were boys of a similar age also being seen off by their parents sitting alongside and opposite and like me putting on a brave face. On arrival at Cosford, after changing at Wolverhampton, we were met by friendly RAF staff who marched us to a line of wooden huts where we were to be accommodated. Following a nervous night, we were fed and watered then taken to an education block to sit the entrance exam. Years of fear of anything educational came flooding back to me and I felt so vulnerable.

The moment finally arrived when we were instructed to open the sealed exam papers. On seeing the contents, I felt a sense of relief to discover the questions on the papers were familiar to me and all I had to do was to keep my nerve and remember what had been said to me by Mr Price. What I had not seen or rehearsed before were the problem-solving puzzles that were to follow which formed part of the examination. Somehow, I managed to complete all the tasks that were set, so I kept my fingers crossed that what I had done would be enough to pass. To complete the process, which lasted two full days, I had to undergo a physical examination, which I undertook without any problems.

My return journey to London was filled with an air of quiet confidence, despite having to muddle through the best I could when sitting the English papers. It was a matter of a few weeks, which at the time seemed a lifetime, before the long-awaited

OHMS brown envelope arrived through the letterbox. To my relief I had passed.

Following a few formalities, I was notified of a course place (14 Entry) commencing a few weeks later. Following preparation and packing, and not without a certain amount of apprehension, the date of my departure duly arrived. Armed with my obligatory tomato and egg sandwiches I said goodbye to Mum, and Dad escorted me to Waterloo Station where we said our emotional goodbyes and I joined a variety of familiar faces whom I recognised from when we attended RAF Cosford. It's amazing in a situation where apprehension is common in a group how quickly bonding takes place.

Unlike my previous arrival at RAF Cosford railway station where we were met by friendly RAF staff, this time the platform was lined with RAF Police dog handlers who compared to us were huge and looked fearful in their tall service headdress with the peaks of their hats lying flat against their nose. This together with their white webbing belt, shoulder straps and gun holster created a reception party that none of us were prepared for. Accompanied by the incessant barking of the police dogs, they shouted out their orders to the coach loads of new recruits. My lasting memory of this frightening experience I can only liken to the Nazis marshalling prisoners into a prisoner of war camp, not helped by the fact that everyone had difficulty in keeping in line carrying heavy suitcases.

We arrived at our barracks terrified and confused to what was, and still is known to this day as, the Fulton Block. We were then ushered and assigned to our 16 to-a-room accommodation where we were shown our bed spaces. On the bed was a set of mug and irons (knife, fork and spoon). From here we were then taken down to the communal mess hall where we were fed and watered.

The following day we were kitted out with our new uniforms and told to pack our civilian clothes, which were to be dispatched home. After many arrival briefs and a swearing in ceremony we were then given our service numbers. We were then placed in small groups and shown by the Drill Instructor (DI) how to look after and iron our new uniforms. This procedure included a DI giving a demonstration on polishing and bulling our shoes and boots to provide a toe cap mirror finish. Shortly after these demonstrations, I came into conflict with one DI because of an issue concerning my boots. Thinking I was religiously following their hints and tips on getting rid of the bobble rough leather finish that existed on new boots and shoes by setting fire to a small amount of polish applied to the boot and rubbed in a circular fashion with the aid of the back of a spoon, I made the fatal mistake of applying too much polish and beeswax in one go. Instead of getting a gentle flame when lit, I produced an inferno, setting fire to the stitching holding the uppers and sole together! To my horror the soles dropped down and parted company with the rest of the boot, which I immediately dropped, burning the highly polished wooden floor next to my bed space. I was duly disciplined and awarded three days of punishment (Jankers), the charge being damaging Government property, not the sort of beginning that I anticipated to kick start my chosen career with the RAF.

CHAPTER 2

Learning Process

The results of the entrance examination and suitability tests governed which aircraft trade that we were to be selected for. To my relief I was to join the Airframe Training Squadron. This meant I was to be trained as a mechanic in aspects of aircraft construction, flying control systems, operational check out and examination procedures. This together with the handling of associated test and maintenance equipment, meant that I was in for a challenging and busy time, not to mention the educational programme that ran parallel providing supporting knowledge in the science behind many of the concepts of mechanical engineering.

The time now arrived when we were split up and assigned to our class groups. We were introduced to our course instructors, who gave us a conducted tour of the training hangers where we were to progress from preliminary basic airframes to the advanced Meteor twin jet fighter aircraft. Throughout this tour he left us in no doubt what was expected of us and the effort we had to put in to keep pace with the mental and physical demands of the course. I was soon to settle back in a classroom environment, the difference this time, being unlike before where the teacher's desk and blackboard represented a barrier of fear I was now being encouraged to interrupt and ask questions. This put me at ease and gave me the confidence to relax into subjects that were new and interesting. There were frequent occasions where we would be taken from the classroom on an individual basis out into the hanger and asked questions or performed a task by examiners. This was valuable as feedback minimising the chance of any nasty surprises when it came to progress examination time.

The preliminary airframe syllabus included the theoretical exercise of using models to explain the principle of safely mooring seaplanes. Getting my young head around tidal movement and strengths and wind direction and speed gave me all sorts of problems but unlike school, where I would have been left behind the rest of the class, I was afforded one-to-one tuition until I came up to the standard required. Reinforcing the classroom lessons, the learning process was backed-up with practical exercises working on live aircraft under careful supervision. My self-confidence grew. From the start we were given notebooks in which we were expected to record lessons and complete in detail later as homework in what free time we had. This would include system operational procedures and, hand drawn in full colour, mechanical component diagrams including complete hydraulic and pneumatic operational flow systems. We handed in these notebooks each week and they were marked for their accuracy and neatness. The results would count towards our end of phase assessment results. We had to balance the pleasure of learning our trade with frequent drill parades and barrack room inspections where it was easy to fall foul of the drill instructors, as there was no compromise in the high standard of smartness and precision in drill movements, but somehow, we got used to it and took it in our stride.

Several periods a week, either morning or afternoon, were set aside for education and this is where my lack of basics in maths and English came home to roost. From the beginning it was obvious that I was in serious trouble as much of the trade subject matter that was talked about in class went straight over my head. Given the difficulties I was coming up against I often questioned how it was that I managed to pass the entrance examination. The only answer I could come up with was that perhaps Mr Price had some insight into the type of questions I would be asked and he had trained me to answer them. Naturally, I was not in a position to know what must have been said behind closed doors by the educational staff, but it was decided that I would be excused some PE lessons and that with some additional evening work I would be allowed to continue in training. My tutor was a Flying Officer Foster who tailored a programme to enable me to catch up. I can't say it was easy and at no stage did I have any ideas of becoming top of the maths or English class, but it was enough to get me through and support the technical work that I was expected to accomplish in the aircraft hangers.

The technical training was conducted under the supervision and guidance of both service and civilian staff headed by a Squadron Commander. Squadron Leader Castle, although strict, was a man you could trust and respect; if you had any worries or concerns, he was someone you could go to for advice. His father figure style of leadership was not confined between 8 a.m. to 5 p.m., as he would frequently be seen walking the corridors of the Fulton Block in the evenings and many times at weekends as well, stopping to talk to and encourage students. His personal touch affected the lives of everyone and he made it his business to get to know you. As a result, he made us all feel that to fail in any way was not just letting yourself down but him as well.

Apart from normal service rules of discipline, we had to comply with the rules concerning smoking, which was strictly forbidden. Our wages were 10/- shillings, equivalent now to 50p per week, of which 7/6d, seven shillings and sixpence (equivalent now 38p) was withheld and put into credits that we got back in a lump sum when we were allowed home at the end of term. Providing you had no stoppages for any breakages or had been fined for any misdemeanours, we took home on average £30 plus ration coupons covering the period of leave. The remaining 2/6d, equivalent now about 12p, had to last us for the week and went towards the purchase of items such as toothpaste, shoe polish, Blanco for our webbing and tins of Brasso. If you were lucky and had some money over, it would go towards a treat of a sticky bun and a cup of tea in the NAAFI! The frequency of going home on leave was based on the pattern of school holidays. Therefore, the longest time away from home occurred during the winter months and was compensated by taking home more credits. I took great pride and pleasure in giving them to my Mum for my keep.

Much of what little free time we had was taken up maintaining the barrack room cleaning with bull nights once a week, during which we would all be assigned cleaning jobs. Just cleaning the bath and sinks were not enough, they had to shine, and the taps had to be polished with no evidence of limescale. The same principle applied to the shower and the floor in the bathroom together with the connecting corridors, which had to be kept spotless. Each bull night was followed by a morning inspection by the DI staff and inspecting officer. Window brass fittings would have to gleam, our personal lockers would have to be dust free inside and out, and our kit folded to a prescribed pattern and size. Over and above this, we would regularly have a full kit inspection where every piece of kit and clothing had to be laid out on the bed in a precisely measured manner that had to conform to the regulation approved photograph. The barrack room's wooden floors were highly polished to an exacting inspection standard; footwear was not permitted to come into contact with it, so we always shuffled along on felt pads. One of the assigned jobs on a domestic night was to polish the heads of the screws securing the floor boards. This proved to be useful because if you had to get up during the night you could navigate your way back to your bed space using them as cat's eyes. Standing next to your bed, which would be positioned in perfect alignment with the other beds, you would await your turn to be inspected knowing full well they would find something wrong. Even cap badges would be removed from your head dress to ensure that it had been polished on the reverse side as well. No detail would be overlooked. Even the needle and thread in its pouch, known as the housewife, had its precise place with the cotton thread folded neatly as per the photograph. Falling short of the standard required on these inspections would result in it all being carried out again the following day or, might even result in an individual being charged and awarded restrictions of some sort, or Jankers.

One of the many traditions ingrained in the system was that the members of the junior entry became bull boys (skivvies) for the senior entry, so not only did you have your own shoes and brasses to clean and polish but theirs as well. If you refused to comply, the custom was to be tried by a kangaroo court of members of the senior entry, and then stripped of all clothing. Under the gaze of class mates the offender would be frog-marched down the long corridor, occasionally being lashed and flicked with folded up towels. Finally, he would be given a cold bath and scrubbed with a bass broom; a similar fate would await anyone who was lacking in personal hygiene, only on that occasion he would be dealt with by his own entry members.

As we progressed further into the course moving on from one aircraft to the other such as the Miles Magister, Tiger Moth, Tempest, Typhoon, Spitfires, Hurricanes and then the jet Meteor we were gaining more experience on the diversity of systems that we would no doubt encounter later in our career. In this controlled environment we would be expected to remove and re-fit the various components and test their performance. Mechanical aircraft systems are notoriously dangerous; therefore, safety was drummed into us from an early stage to enable us to be trusted with the operational check out of hydraulic and pneumatic systems that operated heavy fast-moving components such as landing gear, control surfaces and flaps.

Our First Leave at Home

The time soon came when everybody was looking forward to going on leave for the first time, which was to be Christmas 1951. I remember how excited I was when queueing up in alphabetical order to receive my first credits, ration card and railway warrant. I had not seen such a large sum of money, let alone being given it, and soon my £30 was packed away with the rest of my worldly possessions in my suitcase. The following day we were woken up at our usual time of 06.30hrs to loud music played on the TANOY with the words "you had a good life when you left, ha ha", following which we hurriedly got ready and prepared ourselves for a hearty breakfast. It remained for us to have our obligatory parade and our uniforms inspected. Once we had collected our bags, cases and packed lunches we were marched to Cosford Railway Station. There I boarded my train for Wolverhampton where I was to change, bound for Waterloo and then onward to Forest Hill which seemed a million miles away at the time. My final leg from Waterloo to Forest Hill in Sydenham I remember to this day in graphic detail.

The author pictured outside Waterloo Station while travelling home from RAF Cosford on his first leave, aged 15.

Through the carriage window and passing billowing smoke from the steam engine in front I could see the passing sights of South East London and the extensive bomb damage that still remained, with open spaces where homes once stood. Anderson air raid shelters remained in many back gardens and served as a reminder of times where we spent many uncomfortable day and nights shivering in their cold, damp interiors where, as a precaution against blasts, they had no door, just a Hessian curtain. Contrary to frequent advice, I used to look out through that curtain in the direction of Dulwich where anti-aircraft guns and searchlight emplacements were stationed. I was looking to see if any enemy aircraft were going to be caught in the piercing narrow beams of intense light criss-crossing the dark night sky. It was on those clear nights that I noticed there were three stars close together in a line. I realised that when they were visible, we had little or no bombing raids and they became a source of comfort, so I called them the three wise men who looked after us.

Following any bomb or rocket attacks, of which there were many and very close, the shelter would shake violently followed by a down fall of dust which filtered through gaps in the construction. Ever present was the severe condensation that dripped from the corrugated iron ceiling and walls, forming deep puddles on the floor. Other thoughts on that journey were soon to follow like one unforgettable incident that occurred when, unaccompanied, I left home at Mount Ash Road in Sydenham to go to school, which was always a risk due to the frequent bomber raids. On this particular day I got caught out in the open when suddenly without warning from behind me a German fighter aircraft fast approached low down and opened fire. My saving grace was a nearby large stone gatepost that I hid behind and which shielded me from the fragments of debris flying everywhere. Undeterred I carried on to school where I was to discover a large crowd gathering around the school gates where I was going to meet up with my best friend, Eric Neive, who always waited for me there. At that moment I found out that Eric had been shot and killed by the same low flying German aircraft, later described on the radio as a Me 109 (Bf 109).

Eventually on arriving at Forest Hill, the familiar sights and sounds of the location where I was brought up as a child greeted me. The difference this time was that I was in a smart uniform proudly walking to catch a bus home, while pleasantly reminding myself of how far I had come in a short time and that I was on my way to a promising career, despite a disrupted childhood. Following an emotional welcome home at Dacres Road it felt strange to be able to relax and decide for myself how I was going to spend each day. Although the food back at camp was wholesome and healthy, nothing could compare to Mum's dinners. The time to return to camp soon came around and once again I was being seen off at Waterloo Station by Dad where I boarded the train complete with the usual egg and tomato sandwiches amidst farewells along the platform with other boy entrants and their parents.

Training Intensifies

Once back into the training routine, time was fully taken up with complementary subjects such as sports, the ever-dreaded drill and parades and negotiating assault courses in full combat gear, sometimes under live fire. Cross-country running was the activity that I most disliked. In groups we always started off from the airfield and had to run over two large blister hangers whose grass curved roofs dropped down and were fixed to the ground. This exercise was particularly difficult to do but impossible to avoid because PE directing staff were placed at regular intervals along the course to ensure we didn't cheat. The course would take us through the local village of Tong and Albrighton and back to the airfield.

It soon became my unwise ambition to cheat the cross-country physical training. I set about devising a cunning plan, which if I was to succeed, I would have to somehow avoid missing the roll call at the beginning. After much thought I had worked out a plan but would require the assistance of a reliable accomplice. The idea I had would be to start off as normal in PE kit and have my name ticked off at the roll call outside the Fulton Block. I had identified a large bush which shortly after marching off to the airfield I could jump into, and when the parade was out of view, I would double back into the Fulton Block where a student member on light duties would lock me in my bedside locker and place the key in the webbing on top of the locker. I was hoping that when the room check was carried out by the DI staff I would not be discovered. Following the room check my light duties accomplice would let me out, whereupon I would go in the drying room and do some revising.

On the next occasion of the cross-county run the plan went like clockwork up to the point where I was to be let out of my locker. What I had not foreseen was that my accomplice was to be ordered out of the Fulton Bock before being able to release me, and he was tasked to carry out some light duty work for the DI staff. By the time he was able to get back to me and let me out the cross-country had finished and on opening the locker door I fell out unconscious. The next thing I remember was waking up in hospital where I was kept under observation for a few days because I had suffered partial suffocation. On being discharged from the hospital I was formerly charged with causing a self-inflicted injury and awarded seven days Jankers, which meant more inspections and cleaning work in the evenings. The worst part was having to run around the parade ground several times with a rifle above my head, which was not easy. During this punishment I discovered that the bass broom used for sweeping the cookhouse floor was routinely used for stirring the large vats where some of the left-over food of the day was being converted into the following day's soup. It was a sharp lesson that not only changed my eating preference (a pity, because I have to admit I had previously found the soup very tasty), but a lesson also not to try and put one over on the system again.

As I progressed from the elementary phases of training, the pace of the classroom work and intensity of hands-on experience working on Miles Magisters through to the jet Meteor aircraft increased. Even the notes and coloured diagrams that were set as homework had to be handed in on time to be assessed for accuracy. What we had learned was monitored frequently with a series of progress examinations that consisted of a written theory paper followed by a closely monitored practical task. By this time my maths was up to a reasonable standard and became less of a problem, enabling me to become more relaxed.

Despite my best intentions of not going against the rules, it was common practice taking it in turns having Sunday breakfast brought up to you in the billet from the mess hall on the ground floor by a room-mate as a luxury served in bed, which was forbidden. One fateful Sunday morning when it was my turn to have this rare treat I was sitting up in bed with a full English breakfast on my lap when to my horror Squadron Leader Castle walked into the room. Fortunately, he had his back to me and before he could turn around, I just managed to hide the half-consumed breakfast under the sheet and blankets, supporting their weight the best way I could with my hands so not to ruin the bedding. It did not take a moment when he turned around with the greeting "Good morning Folkes" and moved over to my bed space. I immediately apologised for being in bed and unable to stand to attention. With a friendly and reassuring smile, he sat on the side of the bed and engaged me in a conversation which I had the greatest difficulty in following as he gradually pulled the bedding tighter across me under his weight. I cannot describe the discomfort of an English breakfast fry-up filtering through my pyjamas settling into a gooey mess around where I was sitting. After what seemed to be an eternity, he bid me a cheery farewell and left the room taking no further action. At the time I think I would have preferred to have been charged, but his style of management was more profound and subtle and this was to be an episode which I would not wish to repeat.

Summer Camp

The time soon passed and following a few bouts of end-of-term leave at home, the time for a week's summer camp arrived. We were all kitted out with camping gear and a hamper of food to eat on our long journey by train to Millom, near Barrow-in-Furness in Cumbria. On arrival we set up camp next to the beach and then divided into teams and set about erecting the lines of tents, mess tent and setting up the latrines.

The object to most of our activity whether it was swimming, cleaning or generally getting involved in events, was to organise our own time management. One of these tasks was to climb a small mountain 600 meters high at a distance about 5 miles away called Black Combe. Having been briefed on what daily duties we were expected to carry out, it would be up to each individual to manage and carry out in any order his assigned tasks. Amidst a background of typical summer weather of heavy showers and sunshine, and having prioritised everything that I was assigned, I set about my work for the day. From where I stood Black Combe didn't look too difficult a challenge and as transport was laid on to get there, I decided to leave this climb to last and complete the other assigned duties first. About mid-afternoon I sat off to the base of the mountain climb, which looking up seemed considerably more difficult than I first thought.

The climb consisted of very steep and slippery greenery with a network of narrow well-trodden footpaths winding through heavy rock formations. After much exertion and taking far longer than I had allowed for, I arrived at the top. I quickly scratched my name on a piece of slate-like material and as instructed placed it on the top of a tall pillar of other slate records of having climb up there. By this time, it was getting late so to speed up my descent I slid down the damp slippery slopes, on occasions getting soaking wet by falling into small shallow streams that I had not seen emerging from the side of the mountain. Arriving at the base I was to discover that the last transport had left and I was faced with a challenging 5-mile walk back to camp.

On arrival I was tired and bedraggled and to my surprise my absence had not been realised. After relating my story, I was told to dry off and rest in my tent because I had missed out on getting a meal as all the cooking facilities in the mess tent had been cleaned and put away. To stave off the pangs of hunger I remembered I had in my bag of goodies a bar of Ex-Lax which Mum had sent to me in one of her food parcels. Not thinking I would come to any harm I ate several squares. From here it doesn't take a leap in the imagination to realise what was to ensue. Soon after I was confined to the latrines for the rest of the evening and the best part of the night, drenched in the thunderous rain as the latrines had no roof.

Daybreak the following morning when everyone was queueing up for their breakfast, food was the last thing that was on my mind. It was subsequently decided, in the wisdom of the directing staff, that it would be in my interest and that of everyone else for me to spend the rest of the day in easy reach of the latrines and that I was to consider it my duty for the rest of the day to keep the place clean and tidy.

Airfield Training

On return to our training routine at Cosford we fast approached the stage where all we had learned would soon be put in practice at the airfield using live aircraft. The arrival of the new junior intake of boy entrants arrived with their understandable expressions of apprehension heralded our time to go up in the pecking order of seniority. This effectively meant that what precious little free time we had would now be increased as the new entry would take over our duties as bull boys to the senior entry.

After about one year we were to witness something that we had heard about but never for one moment believed we would see at first hand. This concerned the dramatic punishment not taken lightly, but in exceptional circumstances awarded to anyone who fell foul of rules where there was a zero tolerance by the training authorities. One such rule was theft, which one member of the senior entry had been charged with and found guilty. The method of punishment decided upon for the pupil to receive was to be "drummed out" of the service. This news travelled like wildfire throughout the entire station because this rare, ancient form of punishment was to be witnessed by the entire training staff, both instructors and pupils. The method and procedure in which this very dramatic punishment was to be carried out was proceeded by the buttons and any other badges or insignia on an offenders' uniform being removed and then sewn back on with a single thread of cotton. The date arrived when we were formed up in a hollow square formation on the parade square to witness the punishment to be carried out. Flanked either side by two military policemen the young offender was marched sobbing to the centre of the parade square accompanied by the haunting sound of regular beats from side drums draped in black. Once halted facing the solemn parade, his charge was read out by the Station Commander and the reasons explained for his dismissal from the RAF. With several tugs his buttons, badges and insignia fell to the ground. Once again accompanied by drum beats, he was escorted by the two military policemen to the main gates next to the guardroom, where he was handed over to his parents who were waiting outside. The parade then marched off where we were dismissed in an air of dumb silence.

The day finally arrived where we progressed to being the senior entry, and our airfield training was to begin. Putting together our newly found skills we were able to carry out maintenance servicing on the Lancaster and Wellington aircraft. A new experience was to be trained how to hand swing propellers on a small single engine aircraft such as the Tiger Moth. This was scary to say the least as you have one chance to get it right otherwise you would be hit by the opposite blade as the engine burst into life. On the Wellington aircraft engine, the start-up procedure was also exciting and concentrated the mind because standing as close as 18 inches behind the huge propellers, our job was to hand-pump the fuel line until the engine built up enough revolutions after starting up. The wind with billowing white smoke from the propeller wash and engine exhaust was so strong that you had to grab and hang on to the engine cowling to prevent being blown away. Once the engine revolutions had stabilised, we would make a fast retreat, aided by the prop wash, to the rear of the aircraft.

All too soon our training drew to a conclusion and the frequency of practising for our passing out parade increased. The time arrived for our final exams and practical assessment held over a period of several days. Once all assessments were complete, we were paraded in the hangar to wait anxiously for the results. To my relief I had achieved excellent marks, which meant that I would pass out as a Leading Aircraftman (LAC) becoming a Senior Aircraftman (SAC) after six months in RAF service. The pass out formal dinner was a grand occasion where we were allowed within certain boundaries to let our hair down. Speeches and tributes were made, in some cases comical and less than complimentary comments were directed towards the Drill Instructors. With parents in attendance, after they had been accommodated at Cosford the previous night, they were all seated at the head of the parade ground alongside other dignitaries where we proudly performed our faultless passing out parade. Following introductions between my parents and some of the staff, who were closely responsible for my training, we departed by train for home. The only outstanding thing I needed to know was where I was to be posted to. On this matter I would shortly be notified.

As I was about to find out, the strict discipline under which we were trained, and which was maintained throughout, was for a good reason: It prepared us to be fit for purpose in an operational aircraft environment, sometimes under challenging conditions on which lives would depend.

CHAPTER 3

My First Posting

During my short stay at home on leave I received the familiar OHMS brown envelope. I eagerly opened it to find that my first posting was to the area of my first choice, RAF Thorney Island in Hampshire. This time I was to leave home in the opposite direction, south by train and seen off by Dad with the obligatory egg and tomato sandwiches.

At the age now of seventeen and for the first time free of a training environment, I arrived at my new RAF unit at Thorney Island where what was to happen was now for real. Following arrival procedures, I was formally interviewed by the Senior Technical Officer and told that I would begin my work on second line major servicing, working on Anson aircraft. It did not take long before I was able to put into practice my newly acquired skill of wire splicing aircraft flying control cables that I learned at Cosford. Even at this time it was a dying art, so from the word go I felt I was making a worthwhile contribution to the Squadron's workload. The only problem was that I was under age for my signature to be classified as legal on aircraft documents. The legal age was eighteen years, which meant that any work I carried out had to be checked and over signed by a person of age with qualified experience. My work was nevertheless appreciated and I was soon in demand on other aircraft setting up their undercarriage mechanical systems. This was also to include doing fabric skin repairs to the squadrons Tiger Moth and setting up its flying control system.

The facilities for playing sport were excellent, with groundsmen employed to keep the football and cricket pitches in perfect condition. Having played cricket for my school, it did not take long before I put my name forward for a trial to play cricket for the station team and was lucky to be selected to play on a regular basis. This meant I would travel to other RAF stations and to local universities that were in the same league. I was particularly fortunate as it gave me the opportunity to play alongside or against several well-known County players of the day who were undergoing their national service. My idol at the time was Godfrey Evans who played for both England and Kent as wicket keeper, and whose style of play I used to try and emulate with mixed results. It was the custom for RAF stations at this time to stage an annual sports day where the sports field would be divided into three separate enclosures clearly marked, officers and their ladies, senior NCOs and their wives and airmen and their women. Anyone who without authorisation who strayed beyond these boundaries would be left in no doubt about their place in the class system. This rule also applied to the use of the cricket pavilion during matches, where only officer team members were allowed to use its changing facilities and socialise.

Nine months after my arrival and now a senior aircraftman I was told to report to my Squadron Commander. I turned up nervously, wondering what I may have done wrong. To my surprise I was informed that although the standard of my work was excellent, it had been reported to him that working around the limitations of my signature was proving difficult. He said he was therefore recommending me for a fitters (technician) course and that he had the authority to inform me I would be commencing training at RAF St Athan in South Wales on the next available course. This timing would effectively bring me up to the age of seventeen and a half.

Change in Career Direction

The time soon arrived when I was to start my training, which would take 4 months. Although the content and pace of the course were far from easy, I had little difficulty. Unlike training in the Boy Service, we would now be in adult training have a social life with Friday and Saturday nights being reserved for going out dancing. The popular places to visit where the best bands played were St Athan, Barry, Llantwitt Major, Porthcawl, and Penarth.

What proved to be a problem was having more than one girlfriend at a time, and having to ensure that only one turned up at the same time at the same venue; if two were to turn up I would have to make a quick exit. The only time I sailed too close to the wind was when I had a date with a WRAF PE instructor and took her out to see the film Al Jolson Story in Cardiff. Following a few drinks afterwards we decided to head back to camp. Unfortunately, we left it too late and missed catching the last bus that would take us back to east side of camp where the accommodation quarters were, including the WRAF block that was strictly out of bounds to men.

By this time having to find a taxi, it became obvious that we were going to be late in getting back in time for my friend to book in, and that now trouble loomed because there was no distinction between the female ranks concerning the time they had to back in their rooms at night. Our arrival back was long past that time, and the problem now was the front entrance door was locked. Getting her back inside without being seen was not going to be easy. On close examination I was fortunate to find a corridor window that I could open and climbed inside with the intention of quietly opening the main door to let her in. To my complete horror and surprise, the corridor lights came on and standing in front of me larger than life was the stout figure of the WRAF warrant officer in her flannelette nightie complete with hairnet. Needless to say that when I could get a word in edgeways I had great difficulty in explaining my presence. The following morning in her office and after a severe talking to, she went on to say that my date had fully supported the reasons I had given leading up to our late arrival. To my relief, she declared that she would take no further action and the matter was closed.

From here on I concentrated on my course work. During our morning and afternoon training refreshment breaks I remember going to the Red Shield Club, which belonged to the YMCA. For the price of one penny, you would get a one-inch-thick doorstep of toast and always on the brew was a huge pot of tea you could help yourself to. Although confident but not complacent, I took the finals, and later was delighted to be told that I had passed all the papers. With the course behind me, I returned to my unit as a junior technician commanding a respectable salary of £5 per week. Now at the age of eighteen years I could not only sign for my own work but I was now qualified to over sign work done by mechanics. On arrival back I was told by my Flight Commander I was to move over to first line servicing, dealing with the preparation of aircraft prior to, and on return from, aircraft sorties. The days where I stared at the Craven "A" billboard on my way to school and wondered if I would ever achieve my modest ambition had now become a reality.

Three weeks to the day after my return to Thorney Island,

someone pointed out to me after breakfast on leaving the airman's mess that my name had appeared on the station routine orders, clipped to the notice board. I was dumbfounded to read that I had been promoted to the substantive rank of corporal, and later told I was the youngest corporal in the aircraft trade. In those days this rank involved more responsibility and privileges than it does now, and at the tender age of just over eighteen years of age I was about to take charge of a servicing team of men who were in the main considerably older than myself. This would present me to face a steep learning curve because my man management experience was zero. This was in an era where both regular airmen and national servicemen worked side-by-side, so one had to tread carefully to get work completed effectively and on time while maintaining a good balance of morale. I must have done something right as I struck up a good rapport with my thirty mixed trade team members without compromising my position.

The role of Thorney Island as the No. 2 Air Navigational School meant that there was a requirement for ground staff to go in support of short detachments to Gibraltar in the Varsity aircraft. These were frequent exercises occurring every month for the benefit of trainee navigators prior to completing their preliminary training. They would then move on to more advanced aircraft at their operation unit. This was my first experience of going abroad, taking in my stride the additional responsibilities of being away from base where I would be expected to make on-the-spot technical decisions to remedy any aircraft defects and get it safely back to base, and preferably on schedule. This routine was to continue with my work fluctuating between day and night shifts, working on either the Varsity or the Anson aircraft, or when on duty crew attend to any visiting aircraft that flew in.

During one warm spring evening in 1955, not having long taken over the shift, I was visited by the duty engineer on his rounds of the hangars. Whilst discussing the work being carried out, we heard the familiar sound of Varsity engines being started. This was not unusual, except that given the serviceability state of the aircraft there was no reason for engines to be running. The next indication that something was wrong was when a Varsity taxied past my office window with one of my engine mechanics at the controls, an Anglo-Indian called Nanik Agnani. Immediately after contacting air traffic control, they arranged for a fire tender to block the main runway by having it positioned halfway down the intersection of an adjoining runway. Despite several attempts over the intercom by the air traffic controller to persuade Agnani to shut down the engines, he refused then released the brakes for take-off. With the aircraft accelerating towards it, the fire tender had no option but to clear the runway. By this time the entire station was alerted to witness what was happening, and to the surprise of everyone Agnani successfully took off, and after several circuits of the airfield to adjust the propeller pitch gearing, he headed out towards the English Channel and then towards the east. It was at this point the duty Flight Commander, Flight Lieutenant Johnny Smiles, took off in a second Varsity in pursuit of Agnani in an attempt to guide him back to base. It was not long before Meteor fighter jets from the nearby airbase at Tangmere were sent up to intercept him. I remember that it was also not long before radio programme broadcasts were being interrupted with the news that an unqualified airman pilot had absconded taking an RAF three-man crew aircraft with him. It was a while after these actual events took place that we discovered Agnani had flown over London at low altitude and managed to lose the perusing aircraft in the fading light of the evening, heading south towards the French coast. There was much speculation as to his intentions, one being that he might be about to commit suicide and crash into Buckingham Palace.

From what I remember being told shortly after this incident, Agnani was not experienced enough to realise the engines were overheating and failed to open the cooling gills, resulting in both engines catching fire. The aircraft crashed in flames onto a row of houses in Onnaing near Valenciennes in Northern France, killing three civilians as well as himself and injuring others on the ground. The result of the subsequent enquiry concluded in part by saying that Agnani was an amateur licenced pilot who on a previous occasion, prior to him being posted to Thorney Island, attempted without authorisation at RAF Chivnor to fly a Chipmunk aircraft that crashed into a tent when taxiing. None of his unstable history was known to us, which was a serious failure in the system.

CHAPTER 4

A Life Changing Experience
Was to Begin

In the closing stages of this summer of 1955, there was an appeal for volunteers to apply to join a Task Force for a posting to South Australia where the British atomic bomb tests were to be carried out. Shortly after I applied, I received notification that my application was successful. In November 1955 I was given posting instructions to arrive at RAF Weston Zoyland in Somerset, where I was to join a Task Force known as 208.5. Our purpose was to play a supporting role in the British nuclear tests in two stages. The first of these was Operation Mosaic conducted off the west coast of Australia at the Monte Bello Islands; the second was Operation Buffalo, to be held at a place called Maralinga in the South Australian desert. The weather that winter was very cold, with icy winds and snow that made getting about very difficult, particularly on the open spaces of the airfield. Once assembled we were briefed on our duties and went through the formality of signing a document concerning the Official Secrets Act that we were given to understand was valid for 50 years. The accommodation we were assigned to was very basic and consisted of twelve-man wooden huts in which there were two coal burners per hut.

The author shortly after arrival at RAF Weston Zoyland, aged 19.

We were divided up into small teams, then briefed on the work that we had to carry out on the Varsity and Canberra aircraft which consisted of installing modification kits that would enable the aircraft to collect and sample radioactive fallout. Friendships between the team members were soon established and as a treat we would visit the local pub to sample the local brew; for the princely sum of 8 pence a pint, we got to like their scrumpy. Huddled around the inviting open fire we used to copy the locals by quenching the tip of the red-hot poker in our drinks to give it added interest.

It used to be the practice afterwards to visit a cafe nearby where we had a delicious and generous portion of gammon, egg and chips. The owner was a kindly and elderly man who we got to know well and who we used to call "pop"; he treated us like sons. It was not unusual for him to invite us to stay a while longer. He would lock up, then when we were all sitting around the table, he would tell us about his war experiences in Burma and Malaya and the health issues he still suffered from. He had contacted malaria, which is something we didn't particularly want to know because we were going to stage through both of these countries on our way to Australia. As you would imagine our work load was planned down to the last detail, not helped though having to work most of the time exposed to the bitterly cold elements out in the open on the aircraft dispersal.

We perhaps could be forgiven for thinking that being at the forefront of a nuclear arms race, mundane issues such as domestic bull nights prior to a barrack room inspection would not be a pressing issue, but Weston Zoyland would prove to be the exception. A problem occurred following our first domestic evening concerning the two coal burning stoves in the centre of the room. The inspecting officer told us in unkind military terms the standard fell well below what was required. The reason given was that the coal in the heavy steel riveted bunkers was dirty. To rectify this, we were given a tin of black gloss paint and informed to empty the bunkers. The bunkers and the coal were to be washed and, when dry, the coal was to be put back with the top layer painted. As a result of having to comply, the surrounding floor once being highly polished, was now covered in coal dust and the whole business of a further inspection was gone through again.

The medical facilities where we were going were expected to be very basic, so we were required before our departure to have a thorough medical examination. This included us having a dental check up in the back of a 3-ton truck where any necessary treatment was carried out straight away. Our route out to Australia was to take us through many countries, so we would require the appropriate injections. For our yellow fever jab we lined up inside a long wooden hut where we could see the injections being administered, but this was also to be accompanied by a blood sample being taken from the fore finger with a needle that looked quite large and judging from the expression of the patients quite painful. To overcome this uncomfortable procedure, I quickly devised a plan by putting my finger on the opened window shelf which was covered in snow. With my finger now so cold it was white, I was then ushered forward only to discover that when stabbed several times with this huge needle no sample was forth coming, after which to my disappointment the forefinger on my other hand was used.

At about the end of February 1956, or perhaps a little later, final preparations were being made, including deciding what spares we might need to take with us. With our personal documentation in place including having been positively vetted, the time for the first aircraft to leave along with its servicing soon arrived. My time to leave was to follow a little while later. The Varsity aircraft that I was assigned to unusually had as its pilot a sergeant named Mayna, not the usual commissioned pilot. Our navigator was a flying officer named Malpas and the radio operator, whose name I cannot remember, was a Flight Lieutenant who by virtue of his rank was to be our captain. The ground crew consisted of myself, Jock Gardener and Charlie Foster. Loaded up with a maximum of 1004 imperial gallons of aviation fuel and a full complement of aircraft spares brought the aircraft up to its maximum all up weight for take-off. Our call sign was "Hotel Papa Charlie 17". With our final briefing completed, the day arrived for our departure; a journey that was anticipated to take approximately one and a half weeks.

I had no pre-conceived thoughts on the political background of the Cold War. All I knew was that we were about to break new ground by being at the forefront of the nuclear arms race of which I and many others knew very little about. Up to this time over the history of conflicts, mankind had conducted warfare in a theatre where you could see and hear your enemy. This was going to be very different as our enemy was poisonous radiation fallout. Not only could this not be seen and heard but, although we were not aware at the time, and despite the lessons of Hiroshima and Nagasaki, its effects would appear decades later as fatal illnesses.

We Embark on Our Long Journey

None of us dwelt on these thoughts as we had this whole new adventure of flying to Australia in front of us, and beyond this lay an uncertain future for us all. Following a lengthy take off we headed for Bordeaux in south west France for our first refuelling stop. After landing and arriving at the dispersal, we were amazed to see one of the French ground crew sitting on top of a fuel bowser smoking. After filling up to our Maximum fuel load we headed south across eastern Portugal, from there across the Mediterranean for our first overnight stop on the north African coast at Algiers. It was my first experience of breathing in the pleasant scented warm air of herbs and spices from the nearby Souk. The next morning following the routine pre-flight checks and an early take off, we headed for Tripoli for refuelling. From there we followed the coastline onward to El Adam a short distance from Tobruk in Libya. Unfortunately, the aircraft developed several technical problems that took three days to rectify, and we took the opportunity to visit the town of Tobruk. Although this was ten years after the war you could not stray off the roads as there were still many unexploded land mines littering the area, evidence of the hard fighting by the British 8th Army desert rats. Nearby was the moving unforgettable sight of both the allied and German war graves that were maintained in pristine condition. In the Army mess hall, we were greeted with friendly banter of, here come the Brill Cream boys, after which as you might expect was followed by beers all round. We spent the last night at the makeshift open-air cinema where freshly cooked camel steaks we on sale. Although fibrous, they were delicious.

The next leg of our journey was to fly direct to what was known then as Habanya, an airfield 18 miles outside Baghdad, Iraq. We already knew from our briefings that there was a worsening of relations between Egypt and the UK over the Suez Canal, so our flight path was mainly over the Mediterranean north of Egypt and onward to Iraq, flying on the limits of our fuel load.

Unbeknown to us the political situation rapidly worsened. The first indication we had of this was when two MIG-15 Egyptian fighters appeared close either side of us. Over the radio we were ordered to follow them, at the same time leaving us in no doubt by hand gestures that if we failed to do so then we would be shot down. Our navigator, Malpas informed us that we were being led to an airfield called Abu Seir (no longer listed on any map) near Aljizah where some pyramids were clearly visible on our descent. After landing, our aircraft was met by armed guards in an open vehicle where it was made obvious for us to follow it into a secure barbed wired compound.

By this time, it had already been decided by our captain and discussed with the rest of the crew that any discussions with our captors were not to include any mention of the nuclear tests, and that our status deliberately put our passports as government officials was a wise forethought. This enabled us to say that the purpose of our journey had no aggressive military intention. After the engines were shut-down we were immediately surrounded by armed troops. On opening the door, we were told to remain where we were, whereupon an Egyptian military officer invited himself on board and demanded that we hand over our passports. After a token refusal by our captain, it was obvious that this officer was in no mood to negotiate and our captain then told us to hand them to him. Being ordered not to secure the aircraft, we were then ordered to leave where we were, and then marched in single file flanked by armed guards to a reception centre where to our surprise we were offered a light refreshment of fruit and a cold drink. On reflection there was never a hint in the Craven "A" advertisement, all those innocent years ago, that I might ever be taken prisoner.

We were divided into two groups, aircrew and ground crew.

Although under constant guard, we were never threatened. In fact, our meals, given the circumstances, could not have been better. There did come a time when we were taken individually into an interview room and for what seemed an eternity questioned over the purpose of our venture. Fortunately, the cover story we had all agreed to stick to was a simple one to defend. They held our passports, and it would be obvious to them that we were bound for Australia, so we had to convince them that our mission was to hold a joint PR exercise with the Royal Australian Air Force at the Pearce airbase near Perth. Following these interviews, we spent our time sitting in a well-furnished lounge and for two nights slept in single rooms. Not having contact with our aircrew, we could only hope that somebody had managed to contact the British authorities.

On the morning of the third day our two groups were brought together. To our relief, we were told we would be able to leave and were handed back our passports. This diversion meant that we were down to our last reserve of fuel, but they permitted me to refuel without any costs involved. With an escort of MIG-15s we took off with some relief on clearing their border, and headed across to Jordan and onwards to Habanya. Flying mainly over desert regions meant that our journey was a bumpy one and as we approached Iraq the air temperature rose hideously high. The only difference opening the air vents made was to let more hot air in. On our approach to the plateau where the airbase was located, we couldn't wait to open the door, only to discover on landing when taxiing that opening the door was like facing hot air from an oven. Following the formality of de-briefing it was not explained to us why we were never contacted and informed of the worsening situation over Suez, where, if we had known we would have taken a safer diversionary route.

It was important that we kept ourselves as healthy as we could due to the high temperatures, by drinking our daily intake of one salt tablet in one pint of water every 20 minutes particularly when working out in the open. By this time, we were encountering lots of mechanical unserviceability both with the engines and due to the high temperature leaks in the hydraulic systems. It is no exaggeration to say that you could easily fry an egg on the wing surfaces. To rectify these mechanical problems, we had no option but to work throughout the high daylight temperatures and into the relative cool of the evenings. At least the toilet facilities were OK and offered some temporary relief from the inhospitable conditions, or so I thought. On one occasion when entering a cubical, I closed the door and to my horror discovered a huge spider draped over the coat hook. Being only inches away from it, and not wishing to disturb it further and risking it falling on the floor, I had no option but to make a hasty escape by climbing over the wall partition and hope that I was not going to find something worse on the other side.

The mechanical defects would take us several days to rectify and take us over our planned stay. During this time, I had yet another encounter with a local beasty. One late night after working on the aircraft I was walking back to 'Hotel Habanya' (actually our tent site) in the dark when I saw what I first thought was a large stone in front of me. As I approached to kick it away, the lights from a nearby taxi shone over it and I realised it was in fact a camel spider the size of a dinner plate by the light also shining underneath it. Being a species of which I was not particularly fond of, I left in a hurry.

The time eventually arrived when all the faults were rectified and engine ground runs completed. We were then ready to leave on our next leg which would take us across Persia (now Iran) to our scheduled overnight stay at Moripore outside Karachi in Pakistan. For a change we had no problems with the flight and on checking in at the hotel we made a short visit to the local suburbs of Karachi. It was here that I got my first experience of what poverty was really like. The open sewers were a sharp reminder why we had to have so many injections prior to leaving home. Pulling at our heart strings were children in increasing numbers, begging for money. One strange event that remains in my memory was seeing a workman digging a ditch. After pushing the spade into the ground with his foot, it then took two other men to lift it out pulling on the end of a piece of rope which was tied to the shaft of the spade.

Our departure took us on a north east heading to India where we developed problems with the radio gear as we approached the airfield in Delhi. On landing we were met by a representative from the British Embassy who arranged our accommodation in one of Delhi's finest hotels. In complete contrast to what we had up to now, we didn't want for anything and had waiters and a personal servant, which strangely I did not feel comfortable with having being self-reliant up to now. On one occasion following dinner and returning to my room, I encountered a very smartly dressed Indian servant many years older than myself who had fallen asleep on duty in the corridor. On hearing my footsteps and realising that I had seen him, he pleaded with me not to report him, which I obviously didn't.

In the morning we would be woken up by the room servant with a refreshing glass of fruit juice. When he opened the curtains, behind which there was a balcony, I was greeted by several vultures resting on the ornate ironwork and totally unperturbed by my surprised reaction. After breakfast it was arranged that an Embassy staff member would give a conducted tour of the area, including a visit to the Red Fort in Delhi where we were to witness its impressive size and shown inside the huge walls where there was a moat surrounding a central island as I remember, from which prisoners were thrown to crocodiles lurking in the moat. After finishing work on the aircraft, I could not resist putting to the test what had been said to me about having clothes tailor made from start to finish and collected the same day. On visiting a local shop and measured for a new shirt, I was told it would be ready for collection that afternoon. On my return with my crew member Charlie Foster, I was shown to a cubicle to try on my new shirt. While I was admiring the quality and fit in the mirror, Charlie grabbed my arm and pulled me outside the cubicle and pointed to the wall inside where I had been standing, inches from my back where I had been standing was a large spider. After paying the bill we made a quick exit.

Several days were to pass where we were; because of all the technical defects, we were well behind our schedule. Eventually we took off for Dum Dum airbase near Calcutta. Once there it was a shock to the system to see such poverty. What I had seen up to now was nothing compared to the desperate plight of some of the young women who lived and died in the same disease-ridden gutters where they were born. Outside our hotel we were confronted with many young mothers holding their babies with unforgettable staring, haunting eyes, a window into their desperate inner feelings of despair. At night they would be seen huddled up asleep in gutters shielding their babies from the comparative cold of the night.

After a further short delay, we took off for Rangoon, now Yangon, Burma. Upon arrival we were drenched in perspiration due to the high humidity. We were daunted by the prospect of having to work out in the open to fix more frustrating faults that had arisen. This situation was not helped by one of the Burmese ground staff who damaged the nose of the aircraft with a ground power set that was attached to assist with our engine start up. The resulting dent was soon rectified with few hammer blows to smooth out the surface finish. This task completed, we took off and headed for Singapore in south Malaya. In complete contrast to where we had come from, we were about to land in the middle of a monsoon, which meant that the windscreen wipers had great difficulty in coping with the torrential rain. With the runway completely awash and with nowhere to which to divert, we landed amidst a spray of water that completely blocked the pilot's and our vision and increased the build-up of condensation on the windscreen. Taxiing was quite a challenge due to lack of vision. Even the direct vision window wasn't much help as it let the driving rain in, so our pilot had to follow directions over the intercom looking through to blurred views of territory we were not familiar with. Eventually we arrived at the dispersal relieved that we had not incurred any damage to the aircraft.

Once settled into our accommodation and organised into a work routine to make our aircraft serviceable, we had some free time on our hands to enjoy a little bit of the Singapore night life. Despite differences in rank, Dougie Malpas, our navigator and I had by this time become good friends. He was a handsome young man, always debonair in appearance, and whenever the opportunity presented itself, he would visit night clubs, something I could not afford to do.

Loaded up with refreshments for our onward journey, we were about to leave the familiar stars of the northern hemisphere including my adopted stars, the three wise men, which by now I knew as the belt of the Constellation of Orion. Of course, I also now knew the real reason the Germans did not carry out their raids on clear starlit nights, because their planes would easily be seen. Set on an easterly heading we took off for a little-known place called Zamboanga in what in those days was the Dutch East Indies, now in the Philippines.

As we approached the end of our journey over the South China Sea, I positioned myself in the bomb aimer's position below to have a bird's eye view as we flew over the breath-taking views of north Borneo with its mountain top blue lagoons and golden domed temples. Our destination was on a small peninsula joined to the eastern side of Mindanao Island, which we knew to have a short runway laying in an east to west direction, leaving very little margin for landing error; one mistake and we would land up in the water. Fortunately, we landed safely and to my relief we had no technical problems to deal with.

We were scheduled for just an overnight stay, and once settled in our primitive accommodation we were able to visit the local village. Our arrival was met with some excitement by the local community of First Nations people. We had been briefed on them before our arrival and told that they were head hunters, which was bit concerning to say the least, but they showed us no aggression. In quite good English it was explained to us by an elder the process of how they carried out the horrific act of actually shrinking the heads down to about two-thirds of their original size. We were also given the reason why they cannibalised the bodies; this was to take on their enemy's brave characteristics. It was known to us that to traffic such souvenirs was illegal, so we politely refused their offer to buy their fruits of war.

When I awoke in the morning, I discovered that the rustic ceiling above my bed was also the home to a nest of small scorpions scurrying about, but thankfully keeping their distance. Just prior to our departure we were presented with a freshly cut whole stick of bananas that took two of us to carry. Now ready to depart, Sergeant Maynar positioned the aircraft as near to the end of the runway as possible to give us the maximum room for take-off. Heading almost due south we set course across the Celebes Sea and numerous small Indonesian islands to Darwin in northern Australia.

On landing we were to encounter our first serious problem when our starboard Olio leg (under carriage suspension) collapsed, requiring us to shut down the engines and be to be towed across the dispersal into a hangar enclosure where it would be safe to be raised on jacks. On leaving the aircraft our first requirement was to report to immigration where we were informed of numerous "dos and do nots". These included not walking around in bare feet because of trapdoor spiders and always checking underneath toilet seats before use for red-back spiders. Following this worrying registration procedure, our priority was to asses any potential damage to the aircraft. We carried spares for most eventualities, but not for this unusual problem, meaning that we had to contact the nearest base that would have Varsity spares. In this case it was Pearce Field near Perth.

The journey taken by the author after arrival to Australia en route to the test site at Maralinga. Other locations mentioned in the book are also indicated.

Later the following day, the spare we needed was flown over to us in a Canberra jet. The aircraft having suffered no further damage, I was able to fit the new component, then following taxiing trials the aircraft was loaded up. After this long delay we departed for the airfield at Broome on the west coast for an overnight refuelling stop. The cross-country flight went without further problems, which meant that we need not place any demands on Broome's limited resources. Following final flight preparations, we began our 1,000-mile journey and headed south to Pearce Field, a Royal Australian Air Force Base near Perth.

We were not long into our journey when we noticed that severe weather conditions lay ahead, a possibility not mentioned in the pre-flight meteorological briefing. The storm clouds were so widespread we had no chance of avoiding them. Shortly we encountered strong head winds that deteriorated into a heavy thunderstorm accompanied by worsening visibility. We had flown in storms before but this time our aircraft was being buffeted violently and increasing our fuel consumption. Just when we thought conditions couldn't get any worse, we were struck with lightning, which damaged the navigational instrument panel, taking off Dougie's wrist watch in the process, remarkably not leaving a mark on him. Communications were further damaged at the same time because the trailing aerial that had been lowered outside the aircraft to increase radio transmission (RT) signal was also hit and rendered useless. The good news was that the integrity of the aircraft structure was not damaged, but the uncomfortable reality was we had no navigational instruments including the compass to work out our position, and no long-range radio communication. Suddenly the feeling of the distance from the safety of my home crossed my mind, as I am sure it did for the rest of the crew.

After what seemed an eternity, visibility improved, allowing our captain to decide that our best bet was to fly at a lower altitude and for us all to look out for prominent ground features Doug might identify on his map that might hold a clue as to our position. By this time fuel was becoming a problem and with fading light questions were being asked about whether we were going to make it to our destination. On the flight deck there was a sense of growing concern etched on the faces of the aircrew. Our options were openly discussed: Should we force land on terrain whose surface could not be relied upon, or with the little light that was left head towards where it was thought the coast was and ditch in the sea? Either way our situation did not look good for our survival and my inner priority turned towards my girlfriend Irene. I wanted her to know that should anything happen to me my last thoughts were with her and that I loved her. I took out my small note book and scribbled down my thoughts to her, then put it in my trouser pocket so if the worst happened it would be found.

Dougie, who was by now sitting in the co-pilot's seat, noticed a discernible glow on the horizon. After a short discussion between the aircrew, it was decided it must be the city of Perth as everywhere else was in comparative darkness. Flying blind, it was now a real gamble between being dangerously low on fuel and operating in low light that we headed towards what was thought to be Pearce Field. Not long after this our captain was able to re-establish radio contact, and with confirmation of our sighting we were directed by air traffic control at Pearce Field to the landing site, where to our relief the runway lights were in full view. By this time our fuel was seriously low. Taking no chances in doing an approach circuit, Malpas went straight in and landed. We were given a "Follow Me" vehicle which took us to our designated dispersal.

A thorough inspection of the storm damage showed that most of the avionics components would have to be replaced. With new compasses fitted, the aircraft would need to be compass swung, meaning it would be required to be moved to a remote area free from buildings and towed in endless circles to calibrate the compass until it and the aircraft structure were magnetically in phase, ensuring complete accuracy in compass headings. Sorting out other minor mechanical problems meant that our journey had yet again to be increased, this time by a further week. The aircraft spares, which up to now we had carried with us, had to be left behind to make the aircraft lighter as the distance on our next leg to Adelaide was 1,400 miles, which was at the limit of our flying range. The implications of having taken so long to fly out meant that we had missed taking part in the initial phase of Operation Mosaic at Monte Bello Islands.

Once ready, we took off with our route taking us mainly over the waters of the Great Australian Bight where after our long journey we eventually landed at the airport just outside Adelaide, this time without any technical problems. After a night's rest and our final pre-flight checks completed, we took off on the long-awaited final leg of our journey 550 miles to Maralinga, on the border of the Great Victorian Desert and the Nullarbor Plain and to be our home for the next six months. On landing we were greeted with cheers and teased with typical service humour for having taken so long, a total of six weeks.

CHAPTER 5

Baptism of Innocence

Once rested we were surprised to discover that the site contractors had failed to complete their work on all road, drainage, and sewage construction and for security reasons made to leave the airfield complex. This led to all military personnel having to complete the outstanding work using the heavy plant equipment that was left behind, a process that took about two weeks to complete. What did not go down well was that despite the whole nuclear test project being British funded, and unlike the Australian troops who were living in well-constructed and insulated huts, we were assigned to live in two-man tents.

Our first experience was for us all to be issued with a "dose rate film badge" that was pinned to our outer clothing, the purpose of which was to record our cumulative radiation levels, which we were told would be entered on our medical records. Shortly after, preparations for the tests got under way and to my surprise I was told by my Squadron Commander that in addition to my normal aircraft work I would be assisting in the radioactive monitoring programme working alongside the boffins (scientists). Further briefings were to follow where I was taken to the forward area where the atomic test explosions were to be carried out.

It was a surreal landscape covered with specialist equipment, including, and placed at strategic distances from the Detonation Zone (DZ), tanks and Swift aircraft. Also included were newly constructed buildings of various shapes and sizes. What was a surprise, even to my untrained eye, was how close to the DZ the trenches were, in which I was told live troops would be placed to witness the immediate effects of a close-by nuclear detonation. These bombs with their detonating mechanism were placed on top of steel towers about 50ft above the ground and to keep the equipment cool as possible the outside surfaces we painted white to reflect the sun's heat. From here I was given basic instructions on how to operate the various radiation monitoring instruments and record their findings.

Once the nuclear test programme proper was underway, I acted as a crew member to operate the "whistling tomcat" as it was called. This gadget contained dry ice that included a probe which when selected protruded into the airstream, known to us the "cold finger". It was this cold finger that would take bomb cloud radiation samples at various distances as the cloud we followed drifted down range from DZ. The Varsity aircraft that we had modified back at Weston Zoyland with this equipment was not able to be pressurised, therefore it leaked like a sieve allowing the contaminated air particles to pass through the small gaps around the door and Direct Vision (DV) windows in front of the pilot. Our pilot for each sortie was a Flight Lieutenant (Ginge) Mathison. I vividly remember on each occasion following these sorties being in the decontamination centre where Ginge would always have difficulty ridding his normally well-groomed moustache of radiation particles. Emerging sometime long after the rest of us had finished, his whiskers looked much the worse for wear after having so many washes.

Most of the work was routine, carrying out the decontamination of aircraft and components following these sorties. Despite all the precautions taken with protective clothing, most of our work was spent gaining access to mechanical systems and requiring at times having to remove our protective gloves to enable the removal of highly radioactive small nuts and bolts. These servicing requirements resulted in frequent visits to the decontamination centre.

Despite the basic living conditions the food was of an excellent standard, my favourite being crayfish which we had on a regular basis. The swarms of flies that gathered inside the tent accompanied by various assortments of beetles who found their way inside the bedding were difficult to contend with. An Aboriginal man had somehow found himself working around the tent site as a cleaner, and for a small fee would sweep out and tidy up around the site. Johnny, as he was known, spoke perfect English.

One day I asked his advice on how I could discourage these beasties from my tent, to which he said to leave it with him. Within a very short time he appeared with a creature that looked prehistoric and more frightening than the unwanted guests I was trying to get rid of. It was a lizard about 8 inches long with pointed horny lumps along its back. Most unusually, it had two heads, one of which was a dummy uppermost, the other underneath was capable of being retracted like a tortoise. This was his defence mechanism against anything that might attack by striking at his dummy head. This being the case, his real head would appear from below and, with the sharp spike on his nose, kill its assailant. Johnny said it was a "mountain devil" and despite its fearful appearance was very friendly.

The good news was that he liked to eat beetles and flies. Tied to my bed post with rope he earned his keep and as a treat in addition to his usual diet of my unwanted pests I would frequently bring him raw mince from the mess, which was most appreciated. He became a good companion, sometimes quite content to relax on my lap, but if startled his sharp claws could be very painful on my bare legs.

A mountain (thorny) devil lizard like the one kept
to control insects and creepy crawlies in the author's tent.

With the tests now under way, one of the ground detonations scheduled to take place about mid-afternoon had, for reasons due to unfavourable weather, been delayed. It wasn't fired until midnight when the weather calmed down and the wind was in the right direction. Following detonation, night was immediately converted into daylight with the clouds evaporating in the wake of the shock wave, exposing a clear night sky.

This delay meant that the radiation monitoring that the other members of the team and I were to carry out at DZ was now rescheduled for 08.00hrs in the morning. Suited up in our white protective clothing, we mounted the land rovers and set off to the forward area, a distance of about 10 miles from camp. I remember coverage of our departure being filmed by some news agency.

On approaching DZ we witnessed what appeared to be grass with a huge mound ominously rising up from what was otherwise a flat desert. As we got closer, on further inspection it became obvious that it was not grass but a thin veneer of glass that due to the intense heat from the explosion had converted the desert sand into a bottle green glass. On dismounting our vehicles, the glass cracked beneath our feet. Walking towards the mound, we were confronted with what was in fact the rim of an enormous crater that I estimated to be about a mile in diameter.

As we approached its base, a member of the team remarked that all we need to see now was the silver fins of a rocket with small green Martians appearing through the bluish haze drifting up from inside the crater. Our task was to monitor and record the intense radiation readings. To add to this fearful experience, we would split up and take further radiation level samples of the damaged equipment including the realistic remains of the dummies inside the tanks and other vehicles. On completion of this task, we would hastily retreat back to base to be checked over with Geiger counters prior to entering the decontamination chamber.

One day an incident took place that was to shock everyone. It concerned my Squadron Commander returning from a routine cloud flying mission. It was the practice when we were on stand down or just having a break from work to relax outside the wooden squadron offices where all the work-related documentation was kept, watching aircraft returning following a routine pattern of completing a circuit of the runway before landing. About 400 yards away was the aircraft control tower, supported on a tall wooden framework that over looked the runway. On this occasion a Varsity flown by my boss to our surprise took a different course. With a full crew on board, he did not line up to land, instead turned and head on towards us and the control tower. The sight of the Varsity rapidly losing height heading directly between us and the control tower was to say the least frightening. He then banked sharply to his right. With his wing tip now below the height of the tower and the entire wing span now almost vertical to the ground he passed through the narrow gap of us and the tower, and then levelled off and climbed, eventually to land safely. This reckless behaviour resulted in him being sent back to the UK, where we were later told that he was court-marshalled and dismissed from the RAF.

Outback Excursions

During what little free time I had I used to venture out on foot into the outback, where on one occasion I heard a rustling sound from some parched looking bushes. Cautiously I moved towards it and was to discover a distressed baby eagle whose nest had been destroyed entangled in the branches. Convinced it had been deserted, I carefully took it back to camp. My immediate priority was to construct a cage in which to keep it safe. This added mouth to feed generated lots of interest and soon help was forthcoming in building it a home. My tent was now beginning to look like a zoo, fortunately with no hostility between my guests. To my relief, and like the lizard, the eagle took readily to being fed by hand, and like my other friend relished raw mince. The task of giving it exercise was shared between friends and it quickly grew to quite a size. It soon became obvious that it required expert attention, particularly with its diet. With the full cooperation from my aircrew friends, I organised for it to be flown to Adelaide on one of the regular aircraft shuttle services and then taken to the zoo where I was told later it lived quite happily.

The author with the baby eagle he rescued from an abandoned nest and looked after with the help of friends.

The now fast-growing baby eagle before it was sent to Adelaide Zoo for more expert care.

During a respite in the tests a few friends and I got permission to organize an expedition to the sea and onward to a place once visited by a 19th century English explorer William Tietkens, and from there inland to visit an Aboriginal community, a round trip of about 300 miles. Loaded up with food and camping gear we headed for the coast where we appreciated being refreshed by the sea air and where it was decided to set up a base camp from where we would venture out to visit the local scenery. Our focus of interest was the opportunity to see at first hand the lifestyle of Aboriginal people living in their traditional setting. Leaving camp early, we passed through some remarkable salt lakes, and in one case on one side of the track the water was a deep blue and opposite it was a blood red, caused, we were told later, by chemicals in the salt deposits.

Marked on our map was the location of a place called Tietkens Well, which we knew very little about. Later we learned it was named after a British explorer and naturalist. William Tietkens was born in England in 1844 and emigrated to Australia in1859. He was the first man to photograph Uluru (Ayers Rock) and Tjuta (Mount Olga). As a naturalist he collected many new species of plant and rock samples, and later became a Fellow of the Royal Geographical Society. He died at Lithgow in New South Wales on 19 April 1915. Tietkens Well was but a short diversion from our planned route. We headed to its location, only to find two sets of four wooden stakes about a foot in height sticking out of the ground which were remains of legs for a bed. Close by was the well, about three feet in diameter. A stone we dropped to determine how deep it was took several seconds before hitting the bottom. The story goes that Tietkens and an Aboriginal companion stayed there on a surveying exercise, and he dug the well for their survival with the knowledge of his companion. Although a small, unremarkable patch of land, to us it represented the determination and bravery of a man driven by a desire to understand nature.

It did not take long before we headed off to find the village of the Aboriginal community. As we got closer to the settlement, we were greeted by what was explained as a hunting party, their bodies and faces painted in traditional cultural markings, representing their status within the hunting group. Seeing them armed with huge spears we nervously introduced ourselves but were greeted in return in a friendly manner and escorted to their settlement close by. On our arrival we were struck by how small their makeshift dwellings were. They consisted of a few leafed branches tied together at the top to form an arch, then covered in leaves and other selected vegetation built high and wide enough to accommodate only two people to sleep in with their dog.

We were surprised to be greeted by a non-Indigenous white man who introduced himself and told us of his work there as a French missionary. We were then invited into his wooden constructed hut that was his home and, as far as we could, see his place of worship. His life was dedicated to bringing the Christian way of life to the Aboriginal people of this community. He told us of an occasion when, after relating to an attentive group of the community about the death of Christ on the cross, he was awoken during the night by, in his words "a large blackfella", who was distressed and crying. The story of Christ's suffering in having to endure so much pain was too much for him. The missionary said this man's reaction alone had made his work worthwhile. Apart from a few Western clothes given to them, the people of this community were self-sufficient. Later outside, we politely refused to share in a meal of a freshly cooked lizard (my lizard friend back at camp would not have forgiven me).

A photograph of the RAF group with people from the Aboriginal community they visited while on a camping expedition in the outback.

The people of the Aboriginal community were by this time eager to show off to us their spear and boomerang throwing skills. One member of our group donated a Benson & Hedges cigarette tin to be used as a target, and paced out 25 yards where he duly placed it on the ground. Excitedly, they took turns in throwing their spears and with incredible accuracy hit the target every time, going straight through the tin. This was followed by one of the elders making a boomerang using only a small hand-held axe to carve it from a piece of wood. To demonstrate its use, he threw the boomerang in such a way causing it to travel in an arc behind a bird, and downing it while in full flight with deadly effect. His skill was remarkable, being used in the context of survival in harmony with nature. Looking back on this moment, it was apparent that the laws of aerodynamics were being intuitively played out in front of our eyes using just an axe and with the skill and wisdom handed down from previous generations to carve an object that could fly so accurately. This one exciting demonstration was a moment in time of an understanding between our very different cultures, bring our two worlds together in appreciation of what it takes, and the skill needed, to make an object fly.

Shortly after taking photographs our short visit was to end. With excitable shouting and waving from all in the community, we headed back to our base camp. It was soon time to return back to the Maralinga village. The journey we had undertaken had given us an experience of a lifetime but now it was time to return to reality.

A member of the Aboriginal community showing traditional hunting skills using spears.

An elder from the Aboriginal community skilfully curving a boomerang with a small hand-held axe that he later demonstrated the use of with deadly accurate effect.

A Lucky Escape

In addition to the ground bursts using the towers, delivering the bomb carried by a V-bomber was also practised. As was the custom the task force would be assembled without protective clothing out in the open 5 miles from the drop zone. Relayed over the intercom from the aircraft as it approached, we could hear the bomb aimer count down to the point of releasing the bomb. On one occasion, the pilot fully opened the throttles before raising the flaps resulting in their mechanical operating rods being forced through the trailing edge of the wings. On the word of command, we would remove our shirts, roll them up, about face from the direction from which the explosion was to take place and cover our eyes with a rolled-up shirt. As experienced by other members present at the time, the intense light would allow you to see your hand and fingers as an x-ray would see them, showing just the bone structure. The accompanying heat was like someone placing a red-hot iron close to your unprotected back, making you flinch. The noise built up to its maximum in less than a second. I can only describe its effect on me as feeling what it must be like to be in a microwave cooking from the inside out, making me shake. Such was its intensity it momentarily took my breath away, leaving me to this day with a slight trembling of the hands.

The damage from each bomb test created a scene at the forward area that looked surreal. The once pristine Swift aircraft were now unrecognisable having been twisted and bent out of their once sleek shape by the vortex of the bomb blast. Tanks were blown from their positions like toys with their turrets and gun barrels detached from their mountings and deposited some distance away. The buildings were in most cases now removed from their foundations, others further away although severely damaged were at least standing.

An important part of the sampling programme was that in addition to the Varsity sorties that were carried out at medium to low levels it was also necessary for radiation samples to be taken closer and at higher altitudes to the point of detonation. This task was allocated to the jet Canberra aircraft, and it was to one of these early sorties that I was assigned to take part having never having flown in a Canberra before. The crew consisted of a young pilot behind whom was the navigator, or in this case I think it was just a boffin. My position was to the right of the pilot sitting on what was known as the rumble seat, which pulled down from the side and was located just rear of the entrance door.

On instructions from behind, it was my job to operate various switches on the modified instrument panel in front of me to start the sampling process. The time came for take-off and as we accelerated down the runway on looking over my right shoulder I could see the ground flashing past, then just as the wheels parted with the runway the pilot took a sharp bank to starboard (right). Seeing the ground so close from this unusual angle and speed left me in no doubt that we were in for an exciting time. It was my first experience from our briefing that on this occasion we were not to follow the cloud as it dispersed down range, but this time our contact arrival with the bomb cloud was to be as it rose upwards after detonation. Our rendezvous with the objective was as I remember was at about 18,000 ft. As planned just prior to the detonation we flew in the opposite direction to DZ. As soon as the moment arrived, and upon seeing the flash, we then turned through 180 degrees and climbed further towards the cloud, which by now was not fully formed into its familiar mushroom shape. From my position I was surprised to see some of the dials and flying instruments fluctuating and reading off their normal scale, which on reflection I now realise must have been caused by an intense electromagnetic pulse (EMP) from the nuclear detonation. Looking down I witnessed the frightening sight of a blazing turbulent inferno comprising of a mixture of black and crimson billowing smoke rapidly rising up towards us. Still climbing and getting closer to our point of contact we encountered the shock wave, which flipped the aircraft over on its back but still climbing; thankfully I was held in by my seat straps. We were now heading in the opposite direction away from the cloud.

Having been so violently deflected from our objective and now unable to continue with sampling the cloud, it became obvious to me that we could never have out climbed the speed at which the cloud was rising, and was a blessing in disguise because I do not think we would have got back in one piece had we continued much further on. Assured by our pilot that he had full control, we returned to base. Following our landing and shutting down the engines, on exiting the aircraft door I was not alone in being surprised at the reception there to greet us, most of them looking at the aircraft. All the paintwork with its identification markings was blistered, with an overall appearance of a whitish blush finish, clearly the result of the intense heat it had been subjected to.

An atomic bomb detonation from Operation Buffalo at Maralinga. The arrow identifies the Canberra aircraft making an attempt to take radioactive samples, in which the author was present to operate the sampling equipment controls.

Death Chamber

Since those times I can only comment on what I and many others experienced during those nuclear tests. The real issue now is: Did those events during the 1950s and the 1960s leave a lasting legacy for those who took part and their families? Just one of these concerns foremost in my mind is the choice of the Varsity aircraft in the first place.

Following the distinguished service of the famous Wellington bomber aircraft, the Vickers-Armstrongs Varsity, a direct descendent, came into RAF service in 1951. Its role was to train navigators, wireless operators, bomb aimers and pilots, many of whom went on to fly in the V bomber force. Its vital statistics by modern standards were quite modest, having twin Bristol Hercules 264 power plants, each consisting of 14 radial cylinders developing 1,950 hp at 10,000 ft, giving it a maximum speed of 288 mph. The Varsity was for many years a workhorse used in training aircrew until it was withdrawn from RAF service in May 1976.

I can confirm that in the 10 years I worked on her as an airframe technician and flew many hours as ground crew in support of navigational training, although the Varsity was at times a technical challenge to keep serviceable, it was always safe to fly. What could not have been foreseen was the political decision to use her as a weapon of war, namely carrying out radiation monitoring for the nuclear tests. There cannot be any doubt that whoever tasked the Varsity for these missions must have known that, unlike its counterpart the twin jet Canberra aircraft, it could not be pressurised. Again, I can speak from personal experience, including the six weeks it took us to fly to Australia in the first place, that the Varsity airframe was draughty due to the many leaks from the outside airstream to all quarters of the airframe. Following each radiation sampling sortie, it would take us a long time under the shower in the decontamination centre to bring down the contamination we had on our skin to what was then considered to be a safe level.

The Varsity sampling missions took place at medium cloud levels where the higher radiation readings were generally found. After each sortie, the external airframe would be hosed down on the outside, but it was never possible to attempt decontaminating the inside, resulting in a build-up of untreated radiation levels that was dangerous to both aircrew and ground staff. It is tragic that after the prominent role that the Varsity once played in the lives of so many people, I now always see her as a death chamber for the many who flew in her and did not survive their Maralinga experience, including, as I was to find out much later, my pilot Flt. Lt. (Ginge) Matherson. There is no doubt in my mind that to fly in her during those times was a potential death sentence, although I doubt this will ever be proved.

Back to Normality

November soon came around, which meant that our tour of duty was at an end. The tests series itself was to continue with a fresh rotation of personnel throughout the rest of the 1950s. For our 208.5 Task Force group what remained now was to make our final preparations to leave, including handing in our radiation dose rate badges. The medics told us that the readings would be entered on our medical records. As for my tent companion, the lizard, I released him to an uncertain future to live surrounded by contaminated soil, and it was now necessary to dismantle and remove some of the tents that had been home.

Without meaning to detract from the serious nature of the events I describe previously, I would like to recount a painful experience I remember happening all too well at this point, which looking back I now see as having a humorous side to it. Due to the corrosive desert conditions, the metal attachments connecting the upper crossed poles proved difficult to remove from the supporting posts when attempting to dismantle the tents. The lads working under my direction were messing about not making any progress in being able to separate them. Not to be defeated, and in my wisdom thinking "If you want a job done, you have to do it yourself", I placed a long steel crowbar between one of the stubborn fittings then forced it down while taking a firm step forward.

I was to immediately discover following a searing pain in my left foot that I had stepped on a plank of wood with a large protruding nail that had penetrated through my boot and foot, and was now visible at the top of my boot! After some initial concern, and typical service humour, it was suggested that I be taken to the medical centre. To make things worse the plank was about 5ft long and I was firmly attached to it halfway down, making any attempt to fit in the Land Rover impossible. Further pain was to follow when it was decided that the only solution was to shorten the plank by sawing it off in front and behind my foot!

On arrival at sick quarters, I was unceremoniously carried into the waiting area where I was surprised to see the director of the nuclear trials, Sir William Penney; he wasted no time in asking something along the lines of, what on Earth I had done. His reply to my answer, in typical old-school fashion of the establishment, took the form of "You bloody fool", followed by the remark that I looked like a slapstick clown, or words to that effect. Not very sympathetic in making light of my situation, but looking back I can see how it must have looked. Shortly after this, sympathy was not forthcoming from the Medical Officer, either, who questioned why I came to the medical centre to remove a plank from my foot.

At this point I was just hoping I could be given help in the form of an anaesthetic before it could be removed but, due to the nature of my injury, he was unable to give any anaesthetic until my foot was free from the boot. Accompanied by a medic who firmly held me in a chair, the Medical Officer swiftly pulled the plank away from my foot. After being given an injection for tetanus and an X-ray taken, it was discovered to my relief that the nail had fortunately passed right through the flesh without coming into contact with anything that would cause any lasting damage.

It had been rumoured at the time that our Varsity aircraft and associated equipment were to remain behind due to their contamination, eventually to be buried somewhere within the Maralinga testing site. I was to later learn, however, that only the Swift aircraft were disposed of in this way. Finally leaving behind what had been a dramatic experience, we were flown to Adelaide where we rested up for a few days, giving us the opportunity to visit the local beaches and get re-acquainted with civilisation by going to a few dances. Shortly after, the time of our departure arrived and we were taken to the Adelaide docks where waiting before us was the passenger ship Orontes. Once on board we were shown to our cabins, which by comparison to what we were used to, were sheer luxury. Quickly unpacking, we assembled on deck along the handrails with the other passengers, the vast majority civilians. It was an emotional scene of people crying and waving goodbye to their loved ones. From ship to shore there was a curtain of colourful ribbon streamers and ticker tape falling heavily like snow. Finally, with the sound of the ship's horn, the mooring ropes and gangways were removed from the quayside, allowing the ship to make its slow departure out to sea. The journey took six weeks to Tilbury London, the same time it took my aircraft to fly out to Australia.

On my return I was posted back to Thorney Island, where I soon picked up where I left off, servicing the aircraft carrying out minor and major inspections. The winter of 1957 was accompanied by heavy show falls where keeping the pathways and runways clear was difficult for the motor transport section. One late afternoon a Varsity on a routine navigational training exercise landed and skidded on ice, forcing it off to one side of the runway, where the port wing tip struck a large mound of snow that had been pushed there by the snow ploughs. On impact it swung the aircraft violently and it cartwheeled, spilling gallons of aviation fuel that caught fire that engulfed the aircraft. None of its crew stood a chance of surviving. Among the tragic loss was my friend Dougie Malpas, who flew as our navigator to Australia.

CHAPTER 6

The Latter Years

Since those times my service career covered four home postings, two overseas tours (one in Cyprus and the other in the Persian Gulf) and three years posting at the Ministry Of Defence (mechanical engineering department at the Old Admiralty Building). It was here that I enjoyed walking and relaxing in St James Park during my lunch breaks. On one of these occasions, to my complete surprise, I saw walking towards me my old Squadron Commander from Maralinga. After 13 years since our time in Australia, we instantly recognised each other. We exchanged pleasantries, during which time he said that he now worked in the city. The sensitive issue of the incident that led to his dismissal from the RAF was never mentioned.

Twenty-six years after leaving Australia, in 1982, as supervisor to airfield training at RAF Halton I retired from the Royal Air Force. During that time I had sobering thoughts of when, rather than if, health issues from the Maralinga experience would catch up with me. Several times over this period I enquired if there were any references on my medical records of radiation recordings, only to find that no entries were ever entered.

*The author while on a lunch hour stroll when working
at the MOD in London in the late 1960s.*

Stepping Into Civilian Life

Waking up on the first day as a civilian was for a moment quite a daunting prospect. My service number and 1250 identity card now had no relevance, and for a short while I felt naked and vulnerable. The thought perhaps that I may have become institutionalised didn't help either, but such thoughts soon passed and the priority of finding a new job soon kicked in. It was not long after this that I was to take up a new post as an Instructional Officer at my old unit RAF Halton, teaching aircraft apprentices.

Despite embarking on a new lifestyle, dark thoughts of my past experience as a radiation sampler were by now indelibly imprinted in my mind. I thought about the risks I and many others had taken, not the least those working close up and personal in so many life-threatening situations. One clear evening I was drawn to looking at the moon, which was at its first quarter phase. It was at this moment that my thoughts were taken back to my youth when I adopted the Three Wise Men in the night sky, who in my mind, I thought were protecting me. Seeing the moon through the safety of a telescope eyepiece, in particular its craters, was not too dissimilar an experience. What I could see was a connection between what I had witnessed at the atomic testing site at Maralinga, although on a smaller scale. The Luna surface with its shadowy haunting abstract shapes formed by the craters of shapes and sizes I could only wonder at.

It was for me a light bulb moment that enabled me to confront the demons that had plagued me for all these years, and to turn the negative thoughts I had of mankind's destructive nature, which I had seen at first hand, into a positive. Seeing the forces of nature that has shaped, and still continues to shape, the Moon and beyond, my life was to follow along a parallel route to my day job by becoming an amateur astronomer.

Throughout the 1980s my interest in astronomy deepened and I joined the local Aylesbury Astronomical Society (AAS) where I attended regular monthly meetings. Weather permitting, I enjoyed the Saturday evenings learning and looking through telescopes, viewing the spectacular sights of the night sky at the observatory site on one of the hilltops near where I lived. It was about this time in the mid-80s that my young son, Stuart, was also beginning to develop his childhood interest in astronomy and loved attending all the lectures that were given by local members of the Society. It did not take long before he became friends with people much older than himself, all of whom, and without exception, thought highly of him.

Throughout the remainder of the 1980s Stuart's interest went from strength to strength and he would accompany me to the annual Herschel lectures in Bath, Somerset. Here I met up with the late Sir Patrick Moore whose liking was, prior to going to the lecture hall, to have beans on toast at a nearby cafe. Later at a local venue in Aylesbury I received on behalf of the AAS a 16-inch telescope reflecting mirror from Sir Patrick. I was later to become chairman of the Society and eventually honoured by being given the position of Honorary Life President. In the late 1980s I was invited by the Astronomer Royal to attend several meetings in London in the company of the late Dr Heather Cooper. He outlined his wish to combine aspects of astronomy into the National Curriculum for schools. From there Heather and I worked together in formulating practical tasks that would provide students with meaningful practical tasks that would give relevance to their maths and science exercises. This system was eventually adopted. Sometime later my name was put forward to the Royal Astronomical Society (RAS) to become a Fellow of the RAS, which was subsequently confirmed.

During the latter stages of the 1980s and into the 1990s I was invited many times to give talk as a guest speaker. On one such occasion during a National Astronomy Science week, I was invited to talk at a meeting in Stratford Upon Avon. Having prepared in the usual way of producing my own slides, on arrival I was to discover that the lecture halls overhead and film projector, despite several attempts, failed to work. Undaunted, dressed in my pin striped suit and armed only with a black chalk board that reached along the entire length of the stage, and with a handful of coloured chalks, I began my talk on the development of the Solar System. I was now committed to talk to a full house and illustrating in chalk on the blackboard what I previously had on my slides.

At the end of my talk, having covered every square inch of the chalk board, my once pristine suit was now covered in multi-coloured chalk dust. My remaining task now was to ask: "Are there any questions"? To my surprise a lady at the back of the audience stood up and asked me: "Do you believe in God"? After what seemed to me an eternity, I was relieved to remember the words from a book written by the English astronomer Sir Fred Hoyle that went something like: Forget the name God and think of it being a cosmic intelligence that is tinkering around in the background. I answered her with my recollection of this quotation and with a silent sigh of relief, she accepted my reply.

The author (leftmost) being presented with a 16-inch telescope mirror on behalf of the Aylesbury Astronomical Society by Sir Patrick Moore (right) in the late 1980s.

The Moon

Throughout this period, I experienced a growing interest in the importance of the Moon to both the Earth's rotational stability and of human existence, which led me to enquire further into the Moon's formation history. It is generally accepted that at some point about four and a half billion years ago across the void of space interplanetary debris was swirling around the newly formed Sun. This cosmic dance, choreographed by gravity, most likely caused a planet sized object to collide off-centre at low relative velocity with what we know today as Earth, resulting in an ejected disc of material surrounding the newly forming Earth. This material proceeded to cool and coalesce into what we see today as our Moon. During this collision, the cores of these two bodies combined, the result being that our planet Earth has a larger liquid iron core that we would normally expect it to have. Fortunately for us, this enhanced core acts to provide a magnetic shield, protecting us from the harmful effects of the Solar Winds. Without this shield, our benign climate would become a nightmare of extreme temperatures with -170 degrees Celsius by night and +130 degrees Celsius by day. By contrast, sometime in the past once having land and lakes, and for reasons not fully understood, the weak magnetic field of Mars switched off and potential for life was destroyed.

The gravitational effect of the Moon is dominant in a way that holds the axial tilt of the Earth steady at 23.5 degrees, by acting as a counter balance. On the next occasion you admire the beauty of our nearest celestial neighbour, spare a thought for its unseen power because it provides us with a tidal system. Intertidal zones were key to enabling life to gain a foothold on land nearly 400 million years ago, to transition from fish to the tetrapod species being a group to which humans belong to. There is evidence of fossilised tracks suggesting some form of amphibious life was undertaking periodic excursions between the intertidal flat zones. The first fossilised transition-type creature to be found is called Ichthyostega, and with a more recent fossil discovery being that of a slightly earlier and more primitive creature called Tiktaalik.

The Moon has given us a protective shield against the harmful effects of the constant Solar winds. Its presence allows Earth a steady rotation around its tilted axis, providing us with the four seasons. Unlike Mars and Venus, whose crusts are land locked, Earth's past fortunate collision has provided it with a thinner crust creating plate tectonics and continental drift, thus permitting the migration of species into a landscape that we see today.

Moon Map

From my observation of the moonscape and the impression that it made on me, I drew a comparison between what I had seen at Maralinga and the Lunar surface. It was at this point that I decided to embark on the ambitious project of mapping the Moon's surface, which was to take me 10 years. My first job was to design and build a telescope that would give me the necessary depth of detail, and for this I built a 10-inch open tube reflecting telescope with a focal ratio of 5. The critical path to give me the scale that I needed was determined by the size of a large desk I had in the spare room, being 5 ft in width and 3 ft deep. This told me that the size of the map must not exceed 5 ft in diameter, and the scale would have to be somewhere in the region of 1:2400,000.

The 10-inch Newtonian telescope designed and built by the author for the purpose of drawing the lunar surface in detail.

What I wanted to capture was the tantalising view of the lunar surface at the point where there is a sharp division between day and night, known as the terminator. For this I had to create an imaginary sunrise over each surface feature that would highlight the shadows. To achieve this, I employed the finest Indian ink pens using the technique of fine dots, or pixels, that varied where required to enhance the density per square inch to create a 3-D effect. This effort, as you might imagine, depended on the weather, but when conditions permitted, I would photograph and draw the images and convert them on to the map. The accuracy of detailed measurement for this was possible by using a micrometer attached to the eye piece. The entire project took nearly 10 years to complete but the effort was well worth the result.

Throughout its history, the Moon has been bombarded with meteorites having velocities calculated between 13 to 18 km/sec, resulting in an impact pressure of 10 kbar or 150,000 psi. One such example of so many impressive impact craters is Copernicus located at 9.7 N 20.0 W, named after the famous 15th century Polish Father of astronomy Nicolaus Copernicus. This crater has a ring mountain 93 km in diameter and 3,760 m deep with terraced walls, a relatively flat floor and a group of central peaks that have heights up to 1,200 m; the height of the wall above the surrounding terrain is 900 m. The magnificence of this impressive feature, created by the forces of nature, comes in to view on Luna day 17 of the Moon's monthly phased cycle.

These thoughts led me to the idea of sharing these observations and drawings with others who are interested, by writing a book to enable the reader to benefit from the same experience and wonder as I have done. It is with much gratitude to my wife Irene's typing skills, and to Stuart my son for his help in formulating the detailed Libration charts, my first book *Observing the Moon* was first published in 2003 (Collins: ISBN 0 00 715431 3). This provided me with the opportunity to draw a range of Luna features in greater detail, such as the crater Copernicus shown in the accompanying picture, which I could not include on the full-size map due to the limitation of scale.

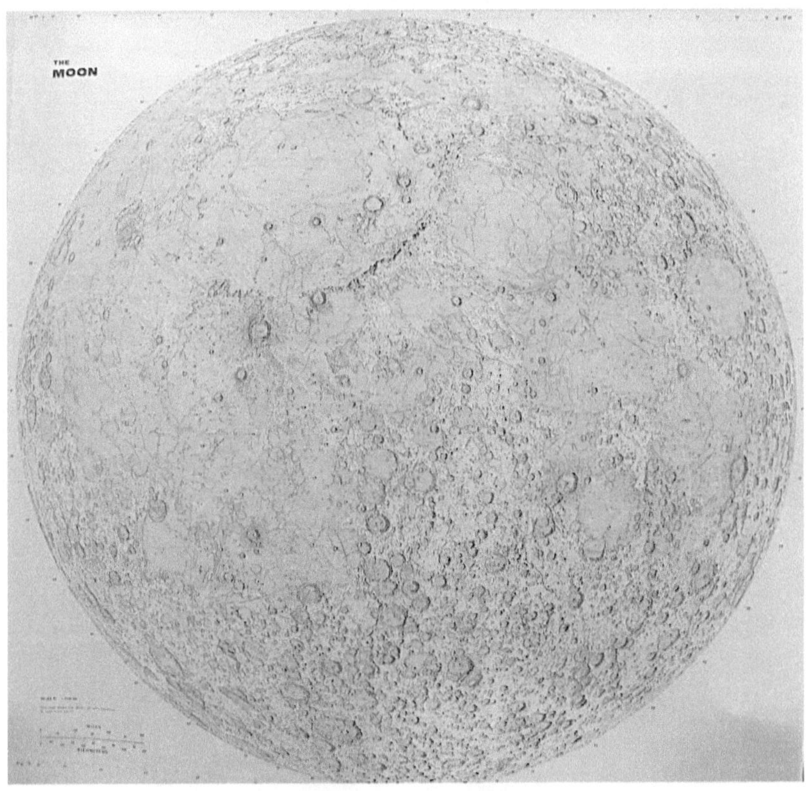

*The final Moon map of diameter 5ft
painstakingly drawn by the author over several years.*

No one situation governs the final shape and form of impact craters, as these are borne out by chance encounters between the impactor and its final destination. Generally, there are four conditions that determine the resultant size and shape of the crater. Three are: The size of the impactor, its velocity and its angle of trajectory. The fourth is the surface composition with which it comes into contact. The physics behind impact cratering is well known, but one can only speculate on some of the complex structures seen to exist within crater formations such as Copernicus.

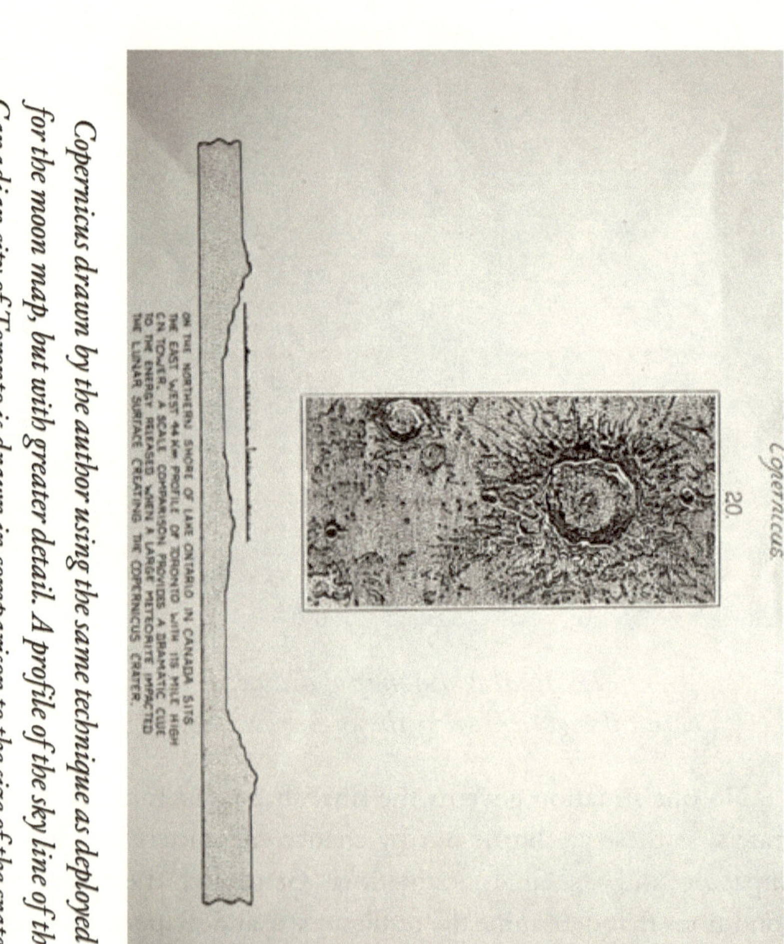

Copernicus drawn by the author using the same technique as deployed for the moon map, but with greater detail. A profile of the sky line of the Canadian city of Toronto is drawn in comparison to the size of the crater.

If only it were possible to be up close to some of these structures, we would look in awe and be overwhelmed by their imposing presence. The reality is that observing is the nearest we are going to get. If we can interpret their form from the comfort of the eyepiece, and comprehend their dimensions, we would be rewarded with a privileged insight in what lies beyond our own planet Earth.

Here on Earth evidence of any impact cratering at the same time as the Moon has all but disappeared, with only a few examples remaining. This is due to natural Earth processes over many millions of years such as volcanism, subduction, continental drift, weathering effects from our protective atmosphere and, of course, the introduction of life in in its various forms. All these factors have modified the Earth's surface to how we see it today. Although volcanic activity played an important part in the Moon's early history, it is the meteor impacts that have been the dominant factor in shaping its surface and character.

From this one experience of recording nature's gift for Earth's survival, the Moon, what I have come to appreciate in my life time is the power that nature has to create an environment and the conditions in which life can grow and flourish. In opposition to this, it has been my personal experience during the Cold War period, and since, that man's attempts to develop high energy forces has one purpose only, to destroy life that has taken millions of years to evolve. Having witnessed at first-hand this destructive power developed by mankind, and if the delicate cosmic dance of natural forces between our two heavenly bodies is to remain intact, the intention of man to return to the Luna surface must carry in its payload a warning never to use it as a base for military purposes.

John Folkes

The Wider Picture

During this time in 1983, the British Nuclear Test Veterans Association (BNTVA) was formed. Its objective was to gain recognition and restitution for test veterans who took part in the British and American nuclear tests at the Montebello Islands, Christmas Island, Maiden Island, and both Maralinga and Emu Field in South Australia. By this time rumours were filtering through of the effects of the radiation that was absorbed by the thousands of servicemen who had taken part. Also of considerable concern was the knowledge that the ensuing illnesses could jump a generation, resulting in dreadful abnormalities in many children. To this day, the work of the BNTVA has fallen on deaf ears with the British Government. I joined this organisation to hear of any progress that might be made in formally recognising its veterans.

Fast forwarding to December 2013, I was asked by the committee of the Boy Entrants association, of which I was also a member, to write an article for their magazine about my experiences during the 1956 nuclear tests. To my surprise, following its publication I received a call from Elizabeth Tynan, an associate Professor at the James Cook University Graduate Research School in Queensland, Australia. At this time, she was researching material for her book, in which she was writing in detailed technical scientific terms about the nuclear tests at Maralinga, and was interested to hear my side of the story. Since that time, we have become good friends, and her much acclaimed book *Atomic Thunder* has been published. *Atomic Thunder* will take its reader beyond my expertise to explain the extent of the damage done to the Indigenous communities. Her book also looks at why these events are still to this day shrouded in mystery and explains the political landscape in force at that time.

CHAPTER 7

Further Thoughts and Reflection

Since those times, nearly seventy years ago, I can only comment on what I and many others experienced during that extraordinary time, and my story would not be complete without further commenting on the serious issues that would follow in its wake.

The uncomfortable truth now is that there can be little doubt that the atomic and hydrogen bomb tests during that time have, for many who took part, created a lasting and cruel legacy of suffering. Not the least of this is that some of the men, thought to be Canadian, who were put in the trenches at the forward area were told that their first exposure to the radiation from the bombs would give them a form of 'immunity' from any subsequent radiation exposure.

During this part of our military history, we shamefully left scattered 22.2kg of plutonium around the Maralinga test site known as Taranaki. The tests there used conventional explosives to blow-up simulated warheads, containing a long-lasting form of plutonium with a half-life of 24,000 years. Its extreme persistence of radiation and the threat of cancer by inhaling small particles in the dust, that is ever present, makes it especially dangerous. This is a story that has conveniently been swept under the carpet during a news blackout. This includes most ordinary Australian citizens who, thanks to Professor Elizabeth Tynan, are now becoming enlightened as to what went on without their knowledge or approval.

Although angered, but not blinded by subsequent events now coming to light, and as a survivor, I feel it is my duty and responsibility to speak out. It seems to me, and to the dwindling remainder of survivors, all of whom endured the most difficult and dangerous of circumstances, that our unwavering dedication of service has been betrayed and conveniently forgotten.

For the first time in military history, we were up against a Cold War enemy of poisonous radiation that was invisible to all our natural senses. The many deaths that were to follow, unlike conventional warfare, did not occur on the battle field, but went unrecorded and unrecognised later at home, whose only record was an empty chair left at the head of the table, looked upon in *silent tears* by the broken lives of their loved ones.

The nature of the radiation and its effect that it had on so many, starting with the fragile emotions that many have suffered, reflect the reality of what they had experienced working and living in a violent threatening environment of the ever-present threat of radiation poisoning. This has been completely dismissed by successive British governing bodies, whose only excuse is that there is only a causal link to the form of plutonium we were subjected to and other forms of cancers such as leukaemia. As a consequence, and despite compelling evidence submitted by the BNTVA on several occasions to the Government's Advisory Military Sub-Committee, these have been rejected, and therefore they do nothing in support or recognition of those who gave the ultimate, their lives.

Because no one had any back up in terms of medical aftercare, the uncomfortable truth is that many lives could have been saved. Consequently, the identity of the insidious radiation that was responsible for so much suffering was allowed to disappear out of sight and mind, together with its victims. There is a question that is never far from my thoughts concerning those aircraft that we were first told would never return to the UK: Despite these aircraft being washed down, did the residual radiation that must have been embedded in countless hidden recesses of the structure and the interior soft furnishings, present a serious health hazard to the unsuspecting and unprotected airman who were subsequently to work and fly in them?

I found evidence supporting this concern from a search of the internet that revealed part of a declassified and partially redacted document issued in 1998 from the MOD AWE research facility (Atomic Weapons Establishment), identifying the fate of some of the 'decontaminated' ex-trials (Operation Buffalo) aircraft. It is only for several Canberra aircraft for which individual details are given, including of their fate, with some being kept in service until 1971. The document also mentions in brief that "all ex-trials Shackleton and Varsity aircraft had been scrapped", presumably by the date of the document, but gives no indication of dates when nor any details of the aircraft involved in the tests. A copy of this document is reproduced in the appendix for reference.

Through the writing of this book, I become acquainted with Simon Batchelor whose late Father, Flying Officer Colin Mitchell Batchelor, flew as navigator on Varsity aircraft at Maralinga during the Buffalo tests. Simon and I have since become friends, and I am indebted to him for a copy of his Father's flight log covering the time period of the tests at Maralinga, providing details of the actual Varsity aircraft used along with their duties. Another search of the internet using the aircraft serial numbers from this flight log that I would most likely to have flown in, indicates that these ex-trials aircraft were, indeed, kept in service with the RAF long after the tests until the late 1960s and early 1970s. One Varsity in particular, WL629, appears to fit in very well with my memory of the events I experienced of the sampling missions, remaining in operational service until 1974, and subsequently being scrapped. The accompanying photograph of WL629 I discovered during the search of the internet was taken some time after its return to the UK from service at Maralinga. The aircraft is shown fitted with orange dayglow adhesive plastic stripes to enhance visibility as an anti-collision measure. These were not fitted to Varsity aircraft during their time at Maralinga.

This period of the Cold War has obviously been an embarrassment to both the MOD and Government, notwithstanding the appalling legacy of the untreated contaminated landmass and territory of the Aboriginal peoples that we left behind. At the annual conference of the BNTVA in August 2021, I put the question to Professor Robert (Bo) Jacobs, Hiroshima City University, live from Japan: "Were we guinea pigs throughout these tests?", to which he replied "No", and qualified his answer by saying "otherwise we would have been followed up with heath checks". I put the same question to Professor Matthew Wiseman, University of Ottawa, Canada, who gave me the same answer. What was of significance was that they were both in agreement that, due to the haste with which the nuclear tests had to be done, we were considered to be acceptable collateral in a race against time.

For me now, when I look back over those difficult times, having served over 30 years as man and boy with pride in a service family in which I put my complete trust, I am now forced to conclude that at the cost of all those who took part, all along we have been an embarrassment. I hope by inviting you into my world you will forgive me for saying that I mean no disrespect to the fallen in past conflicts, but for the present, I am now forced to see much of the military pomp and circumstances in a different light, one of a charade.

For my part now as one of the few survivors having been spared to tell my side of the story, I can say that a sword of Damocles has hung over my head as I have contemplated the real and threatening health concerns I have carried in the back of my mind for so long. In a strange way, telling my story has helped to remove that sword, which has been there for the best part of a lifetime. Those responsible still have a duty of care to live up to, but the feeling of having been deceived and betrayed will unfortunately always remain with me.

Despite a precarious start to my life in having survived the bombing of the Second World War and the nuclear tests of the 1950s, life since then has been much kinder. I realise now looking back during the time when we had a narrow escape trapped in a tropical storm flying from Broome to Pearce Field, that was a pivotal moment in my life where my thoughts turned to Irene, and I wrote down what she meant to me and how much I wanted to spend the rest of my life with her should I survive. In 1959 Irene and I married and had two sons Steven and Stuart. Throughout the rest of my long military service, Irene never wavered in her devotion to her family and as a service wife gave me 100% support. Following a marriage that lasted for over 50 years, in 2011 Irene sadly passed away. In 2013 I married Margaret, to whom I am most grateful for providing her understanding and moral support to me as I have dredged up dark and painful memories from my months at Maralinga.

The RAF Vickers Varsity T. Mk 1 aircraft. The foreground aircraft, WL629, is most likely to be the actual one flown in by the author for radiological sampling survey missions after the second test of Operation Buffalo. Image courtesy of BAE SYSTEMS.

CHAPTER 8

A Lasting Legacy

Including a thorough review of scholarly articles and governmental reports that add to the debate concerning adverse health effects experienced by serviceman and civilian contractors involved with the nuclear tests, caused by exposure to radiation they may have received, is beyond the scope of this book. However, and due to the importance of this contested legacy resulting from the nuclear tests, I wish to provide the reader with an outline of the more pertinent material that has been published in this regard and which is freely available. I have listed these resources under Reference Sources.

The first major review to investigate the conduct of the British nuclear tests was by an Australian Royal Commission in 1985 (McClelland, 1985), whose findings were influential in facilitating the clean-up of the Maralinga test site as well as in helping to capture the attention of the wider public regarding the important issues concerning the tests. Another, later Australian report by Carter, M. et al. (May 2006) concerns the study of dosimetry, mortality and cancer incidence in Australian participants of the tests, following a study commissioned by the British MOD (Muirhead C. R. et al., 2003). This later Australian study found good general agreement in the estimated radiation doses received between British and Australian participants, but that some British participants may have received somewhat higher doses than the most highly exposed Australians. However, the average level of dose received by the British was about half that for the Australians.

In the following four paragraphs I paraphrase the conclusions of a revealing study that considered adverse effects to heath experienced by British and New Zealand nuclear test veterans and their descendants (Roff S. R., 1999):

More than 20,000 British servicemen took part in the atomic tests in Australia and the Pacific, most of them on their National Service, few of them volunteering for the tests, and most in their early twenties and with some still boy soldiers in their teens. These young men were required to participate in the United Kingdom funded nuclear tests in South Australia and Christmas Island during the 1950s and 1960s. Also, 528 members of the New Zealand Navy were present for one series of the tests. Added to these were a Fijian Army contingent, in which an estimated number between 100 and 500 men who also took part. An estimated 16,000 Australian servicemen and civilians were also actively involved at the Maralinga test site and other test ranges. All these men performed a wide range of duties, from highly technical participation in the detonations, to catering, clerical jobs and medical services. A common factor was that we were all required to witness the detonations as a part of their indoctrination for the possibility of a nuclear war.

For this, most of the personnel were required to line up with their backs to the explosion and their hands covering their eyes for a minute or so. They were then allowed to turn round and look at the awesome sight as the mushroom cloud plumed thousands of feet into the sky. Very few wore more than shorts and sandals during their time at the tests. Only those who were thought to be at risk from radiation injury were issued with protective clothing and dose rate badges. The UK government was sure that troops — most of whom were standing out in the open 20 km from the point of detonation, and some of them witnessing 25 nuclear bomb blasts in as many weeks at Christmas Island — would not be harmed.

The Ministry Of Defence still routinely issues a document to nuclear veterans who feel that their illnesses since that time were caused by the radiation they encountered when they were young men, which states: "The background radiation dose that was received by civilians and members of H M Forces serving at or off Christmas Island in the years 1956 to 1963 was only about 35% of what they would have received on average had they remained, for that period in their lives, in the United Kingdom, this is, some 100 micro sieverts per calendar month less at Christmas Island than in the United Kingdom (1 million micro sieverts is equal to 1 sievert)". This sanguine view of the health burden borne by nuclear veterans and their families is not borne out by data reported in the study of the health outcomes of the 2,500 men (UK, 238 New Zealand and Fijians) on whose data are available to present researchers. Some 30% of these men in this study sample have already died, mostly in their fifties, two-thirds of them from cancers that are pensionable in the United States as presumptively radiogenic among nuclear veterans. About one-in-seven of the men in the sample of 1,014, who responded to the questionnaire circulated in 1997, did not father any children after they returned from the weapons tests.

Among the nearly 5,000 children and grandchildren of this group of more than a thousand veterans, there are 26 cases of spina bifida alone, more than five times the usual rate for live births in the United Kingdom. Nearly half the problems among the offspring of the nuclear weapons test veterans reported in this study consisted of the same dermatological, musculoskeletal and gastrointestinal conditions from which many of the men have also suffered. Among the 2,261 children of 1,041 veterans more than 200 skeletal abnormalities were reported, including more than 30 cases of short stature and 78 spinal problems mostly being curvature and scoliosis. More than 100 skin conditions were reported, mostly eczema and dermatitis and for the large part described as congenital. Over 50 of the children are already suffering from arthritis and similar conditions, although they were only entering their thirties at the time of the report. Hip deformities were reported for 19 children and kneecap deformities for 14. More than 100 of the veterans' children reported reproductive difficulties with 24 women diagnosed with problems with their ovaries.

These reports of suffering reported in Roff S. R. (1999) do not stop here, because these radiological injuries as seen by these studies can jump a generation, being passed on and repeated in the grandchildren. It now remains to be seen if the Government will finally recognise both that there is a direct link with the radiation poisoning adsorbed by those who took part in the nuclear tests of the 1950s and 1960s, and the diverse radiological diseases that continue to this day, affecting so many innocent lives.

I think it worthy to note, and as a measure of just how reckless some of the political decisions were at that time made from governments at opposite sides of the world, the events that were permitted to take place shortly following my departure from Maralinga. On 22 November 1956 the Australian Olympic Games opened, held in Melbourne, Victoria, some 1,525 km (946 miles) southeast of the Maralinga Testing Site. I was part of the aircrew flying cloud sampling sorties that were carried out just a few months prior to the start of the Olympic Games. These sorties on occasion would take several hours to complete and cover many hundreds of miles. Due to the length of time that has passed since those events I cannot be certain where we actually flew over when following the clouds, and as not officially aircrew I was not given a logbook in which a record of the sorties I took part in would normally be kept. However, there is *no* question that radioactive fallout reached Melbourne and other highly populated regions to the south and east of the Maralinga test site as well as contaminating the Maralinga village from at least one of the tests, as shown in the accompanying picture of the fallout map of the third Operation Buffalo test (McClelland, 1985). It is an open question whether visitors to the Olympic Games, along with residents of those areas affected, were harmed by exposure to the radiation from this fallout.

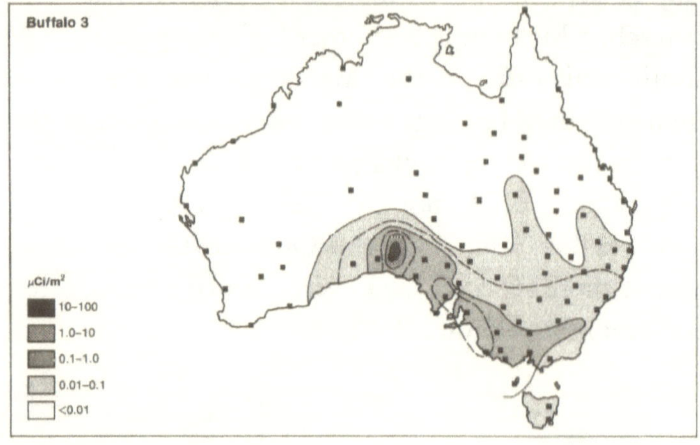

Radioactive fallout map taken from the Australian Royal Commission report (McClelland, 1985) showing the extent of the fallout from the third test of Operation Buffalo over populated areas covering South Australia, Victoria and New South Wales.

In 2017, with a primary focus on the British nuclear test veteran community, the Centre for Health Effects of Radiological and Chemical Agents (hereafter CHRC) was set up in the UK. The CHRC is a centre for research and independent evidence-based information that is available to a broad range of interest groups and hosted by Brunel University of London. I refer the interested reader to the comprehensive information available to view on the CHRC webpages, including general information about the British nuclear test programs, lists of academic articles and information on the effects of radiation exposure, other health and wellbeing studies relating to the British veteran community, and lastly to references therein that support relevant topics I cover in this book. The home webpage for the CHRC website can be found at the end of this book under Further Reading.

A Final Comment

During the frightening experiences, and on many occasions painful ones, my schooling as a young boy resulted in one non-academic qualification that I took with me into the world, their motto: *Service Before Self*. It now remains my hope that the British Government adopt the same principle as a moral compass, and recognise those who put their Country before themselves.

FURTHER READING

i. Website: The Exposure Magazine, published by the Nuclear Communities Charity Fund (NCCF) is a comprehensive and primary source of information for members of the nuclear veterans' community, as well as maintaining a website containing a wealth of news, latest research and other highly relevant on-line resources: URL: https://exposure.press/

ii. Website: Centre for Health Effects of Radiological and Chemical Agents (CHRC). Hosted by Brunel University, London UK. For more specific information concerning health, wellbeing and genetic related research being conducted for the British nuclear test veteran community.URL: https://chrc4veterans.uk/

iii. Book: Tynan, Elizabeth. (2018). *Atomic Thunder*. Pen & Sword Military. ISBN 978-1 52672-757-2.

iv. Book: Tynan, Elizabeth. (2022). *The Secret of Emu Field*. NewSouth Publishing. ISBN 9781742236957.

v. Website: Dr Becky Alexis-Martin. Lecturer in cultural and political geography at Manchester Metropolitan University. A good resource for information on: Social studies of British and International based nuclear test veterans' and their families; the harms caused to Indigenous communities in the places where nuclear tests were conducted; the geographical and social contexts of nuclear warfare. URL: https://www.nucleargeography.com/

REFERENCE SOURCES

i.   McClelland, J. R. (November 1985). *Royal Commission into British Nuclear Tests in Australia*. Vol. III Conclusions and Recommendations. The Parliament of the Commonwealth of Australia.

ii.   Carter, M. et al. (May 2006). *Australian participants in British nuclear tests in Australia*. Vol 1: Dosimetry. Commonwealth of Australia Department of Veterans' Affairs, Canberra. ISBN 1-920720-38-3.

iii.   Muirhead C. R. et al (2003). *Mortality and Cancer Incidence 1952–1998 in UK participants in the UK Atmospheric Nuclear Weapons Tests and Experimental Programmes, National Radiation Protection Board Report NRPB–W27*, National Radiation Protection Board, Chilton (ISBN 0-85951-499-4).

iv.   Roff S. R. (1999). *Mortality and morbidity of members of the British Nuclear Tests Veterans Association and the New Zealand Nuclear Tests Veterans Association and their families*. Med Confl Surviv. Jul-Sep;15 Suppl 1:i-ix, 1-51. PMID: 10467894.

ACKNOWLEDGEMENTS

i. To my Son Dr Stuart Folkes who has helped me through the technical challenges in the preparation of this book, and for his editorial improvements to the manuscript.

ii. To Professor Elizabeth Tynan for her support during what for me have been difficult times.

iii. To the British Nuclear Test Veteran Association for their tireless support to the nuclear Veterans and their families.

iv. My gratitude to Mr Simon Batchelor for making available flight documents belonging to his late Father Flying Officer Colin Mitchell Batchelor who was an RAF Varsity navigator, and for shearing his experiences with me.

v. My thanks to BAE SYSTEMS for granting license to use the image of Vickers Varsity WL629 in this book for which they hold copyright.

APPENDIX

Reproduced here for reference are two pages from part of a declassified document obtained from the internet. Dated March 1998, and originating from the MOD AWE research facility (Atomic Weapons Establishment), it identifies the fate of several 'decontaminated' ex-trials RAF Canberra aircraft in detail, but only makes brief mention of the Varsity aircraft with no details of individual aircraft provided.

Issue 1 March 1998	Aircraft Decontamination at the UK Atmospheric Nuclear Trials	AWE/HPRK/C/REP/HE/98/01 Page 1 of 12

~~RESTRICTED - STAFF~~

APPROVED DOSIMETRY SERVICES

(HEALTH EFFECTS)

Document Reference No: AWE/HPRK/C/REP/HE/98/01

Document Category: C

Issue 1 March 1998

AIRCRAFT DECONTAMINATION AT THE UK
ATMOSPHERIC NUCLEAR TRIALS

Distribution List:

Prepared by   Date 12/5/98

Approved by   Date 13/5/98

~~RESTRICTED - STAFF~~

<table>
<tr><td>Issue 1
March 1998</td><td>Aircraft Decontamination at the UK
Atmospheric Nuclear Trials</td><td>AWE/HPRK/C/REP/HE/98/01
Appendix D Page 1 of 1</td></tr>
</table>

APPENDIX D

Fate of ex-Trials Aircraft

Canberra B6	WH962	To RAF Catterick for firefighting	31/9/71
Canberra B6	WH976	To RAF Catterick for firefighting	30/1/71

Engines 5799/646567 (to 27/7/56), 1583/657072 (from 27/7/56), 1532/657041

Canberra B6	WH978	Struck Off Charge and scrapped, Wroughton	31/10/71

Engines 6371/648917, 6366/648912

Canberra B6	WH979	Struck Off Charge and scrapped, Wroughton	31/10/71

Engines 6339/648885, 6336/648882

Canberra B6	WH980	Struck Off Charge and scrapped, Wroughton	31/10/71

Engines 6382/648928, 6372/648918

Canberra B6	WJ754	Sold to BAC Warton	31/11/67

Engines 5621/646389 (to 27/7/56), 6300/648846 (from 27/7/56), 5690/646458

Canberra B6	WJ757	Sold to BAC Warton	31/11/67
Canberra B6	WT206	Struck Off Charge and scrapped, Wroughton	31/10/71
Canberra B6	WT208	Sold to BAC Warton	30/11/67

Source; Record cards held by RAF Museum, Hendon (M59948)

All Valiants were grounded in 1965 and subsequently scrapped at Gaydon, Marham and Wyton, with the sole exception of XD818 which is on display at the RAF Museum, Hendon.

All ex-trials Shackletons and Varsities have been scrapped.

Available worldwide from Amazon
and from all good bookstores

Michael Terence
Publishing

www.mtp.agency

www.facebook.com/mtp.agency

@mtp_agency

www.ingramcontent.com/pod-product-compliance
Lightning Source LLC
Chambersburg PA
CBHW022105050726
47591CB00002B/677